PLANT GROWTH REGULATORS

A Study Guide for Agricultural Pest Control Advisers

Revised Edition

University *of* California
Agriculture and Natural Resources

PUBLICATION 4047
2003

To order or obtain ANR publications and other products, visit the ANR Communication Services online catalog at http:/ /anrcatalog.ucanr.edu or phone 1-800-994-8849. You can also place orders by mail or FAX, or request a printed catalog of our products from

University of California
Agriculture and Natural Resources
Communication Services
1301 S. 46th Street
Building 478-MC 3580
Richmond, CA 94804-4600

Telephone 1-800-994-8849
(510) 665-2195
FAX (510) 665-3427
E-mail: anrcatalog@ucanr.edu

Publication 4047
ISBN-13: 978-1-60107-418-8

POD-3/15-SB/WFS

To simplify information, trade names of products have been used. No endorsement of named or illustrated products is intended, nor is criticism implied of similar products that are not mentioned or illustrated.

Warning on the Use of Chemicals

Pesticides are poisonous. Always read and carefully follow all precautions and safety recommendations given on the container label. Store all chemicals in their original labeled containers in a locked cabinet or shed, away from foods or feeds, and out of the reach of children, unauthorized persons, pets, and livestock.

Recommendations are based on the best information currently available, and treatments based on them should not leave residues exceeding the tolerance established for any particular chemical. Confine chemicals to the area being treated. THE GROWER IS LEGALLY RESPONSIBLE for residues on the grower's crops as well as for problems caused by drift from the grower's property to other properties or crops.

Consult your county agricultural commissioner for correct methods of disposing of leftover spray materials and empty containers. **Never burn pesticide containers.**

PHYTOTOXICITY: Certain chemicals may cause plant injury if used at the wrong stage of plant development or when temperatures are too high. Injury may also result from excessive amounts or the wrong formulation or from mixing incompatible materials. Inert ingredients, such as wetters, spreaders, emulsifiers, diluents, and solvents, can cause plant injury. Since formulations are often changed by manufacturers, it is possible that plant injury may occur, even though no injury was noted in previous seasons.

Printed on recycled paper.

Contents

Preface

ABOUT THIS PUBLICATION

This publication is the official Study Guide for people studying for the California Pest Control Adviser License in the area of plant growth regulators (PGRs). It will also be very useful for anyone interested in learning about the use of plant growth regulators in California. This publication was originally published as UC ANR Leaflet 4047 in the 1970s and was revised slightly in 1993. The 2002 edition represents a major revision of the work.

Prior to initiating this revision of the study guide, a committee of PGR scientists and pest control advisers (PCAs) who work closely with PGRs met to develop a list of knowledge expectations that would serve as an outline for what new PCAs should need to know about PGRs in order to pass the licensing exam in this area. These knowledge expectations are available for your reference at the California Department of Pesticide Regulation Web page (http://www.cdpr.ca.gov), under professional licensing and certification information. Exam questions are derived directly from the knowledge expectations. This Study Guide covers all the information required to pass the exam plus additional information. Use it in combination with the knowledge expectations when you study for your exam.

In chapter 1, we provide you with an overview of plant growth regulators and their uses. Following chapter 1, tables 2 through 6 list information on some common chemicals currently used in California as plant growth regulators. Chapters 2 through 7 relate to specific crops and to specific PGRs used in those crops and how they are best applied. This information, while important, is likely to change over time. Plant growth regulators are registered as pesticides, and the availability of specific materials is likely to change. Always check the current pesticide labels and consult with your county agricultural commissioner for information to confirm pesticide registration status. Also, the Crop Data Management Systems Web site (http://www.cdms.net) is a good source of current label information.

ACKNOWLEDGMENTS

The authors extend special thanks to Emily Blanco for her help compiling the tables and glossaries and for assuring that the knowledge expectations were covered and to Patricia Gouveia, who assisted in assembling the original knowledge expectations. Many of the chapters include information from the previous editions of this publication, and the authors of those earlier chapters are gratefully acknowledged.

——Mary Louise Flint
UC Statewide IPM Program
Technical Editor

1

Plant Growth Regulators: An Introduction

DAVID W. BURGER, Professor and Horticulturist,
Department of Environmental Horticulture, University of California, Davis

Plant growth regulators (PGRs) are organic compounds that when applied in small quantities promote, inhibit, or otherwise modify physiological processes. Table 1 summarizes the myriad plant growth and development processes that are affected by plant growth substances. The importance of plant growth regulators was first recognized in the 1930s and, as a result of extensive research, natural and synthetic compounds have been discovered that are valuable in many agricultural practices. These compounds can extend the geographical area over which a crop can be grown, influence phenotypic expression and maintain or enhance plant or fruit quality, and increase the production, useable harvest, and postharvest preservation of food and other crops. To understand the commercial potential of plant growth regulators, it helps first to become familiar with what is known about naturally occurring and synthetic plant growth substances.

Plant growth substances (sometimes also called hormones, phytohormones, or growth regulators) are naturally occurring organic compounds that are active at low concentrations and promote, inhibit, or alter plant growth and development. Plant growth substances are translocated from a site of synthesis to a site of action that may or may not be in the tissue or organ where they are synthesized. The translocation of plant growth substances can take place in the plant's *xylem* (water- and nutrient-conducting vascular tissue) and *phloem* (photosynthate-conducting vascular tissue). Plant growth substances that are moving throughout the plant can affect many different kinds of organs and tissues having obvious effects on meristems (e.g., shoot and root apical meristems, vascular cambium).

Some growth substances occur naturally as volatile compounds. Examples include ethylene and the methyl esters of jasmonic acid (JA) and salicylic acid (SA) that are transported through the air, a fact that is relevant both to what goes on in nature and to the ways they might be used in agriculture. The precursors of these volatile forms are non-volatile molecules, which, in the case of JA and SA, are also natural growth substances.

In contrast, synthetic plant growth regulators are chemicals that are not found in plants but may have similar growth-regulating properties when applied to seeds, roots, shoots, or flowers. The general name for plant growth substances that regulate growth is *plant growth regulators*. The naturally occurring plant growth substances are discussed below under Plant Growth Substances and

Table 1. Activities of growth substances in plant growth and development.

	Auxins	Gibberellins	Cytokinins	Inhibitors	Ethylene
Cell division		X	X		
Cell wall loosening	X				
Cell enlargement	X		X		
Root initiation	X		X		
Callus formation	X		X		
Xylem formation	X		X		
Protein synthesis	X	X	X		
Stem elongation	X	X			
Lateral bud growth	X		X	X	X
Alpha-amylase release		X			X
Dormancy		X	X	X	X
Juvenility	X	X			
Growth rate	X	X	X	X	
Flower initiation	X	X	X	X	X
Sex determination	X	X	X		X
Fruit set	X	X	X		X
Fruit growth	X	X	X		X
Fruit ripening	X	X	X		X
Tuberization	X	X	X	X	X
Abscission	X	X	X	X	X
Rooting	X	X	X		X
Senescence	X	X	X	X	X
Seed germination		X	X		X

the synthetic plant growth regulators that are either in use or potentially of use are discussed later under Plant Growth Regulators of Agricultural Interest. Plant growth regulators and their uses as mentioned in this chapter are examples only, and are not necessarily registered for those uses in California. Check current registrations before you recommend or use any pesticide. Tables 2 through 6 (following this chapter) give some examples of plant growth regulators that have been used in California.

PLANT GROWTH SUBSTANCES

Plant growth substances are chemical communicators involved in virtually all plant growth and developmental processes. These substances function in roles that concern the essential aspects of agriculture yield and quality in all their forms.

Five major classes of naturally occurring plant growth substances have been identified in higher plants, and it is conceivable that others will be found. The most studied hormones are the *auxins,* characterized primarily by 3-indoleacetic acid (IAA); *cytokinins,* such as zeatin; the *gibberellins* (GAs); *ethylene;* and *growth inhibitors* such as abscisic acid.

Other naturally occurring substances have been found that have some effect on plant growth or development, and these make up the "other" class of growth substances. This class includes brassinosteroids (stimulate stem elongation, inhibit root growth and development, promote ethylene biosynthesis and epinasty), salicylates (reported to inhibit ethylene biosynthesis, reported to inhibit seed germination, reverse the effects of abscisic acid), jasmonates (promote senescence, abscission, tuber formation, fruit ripening, pigment formation, and tendril coiling), and polyamines (have a wide range of effects on plants; they appear to be essential in growth and cell division). Some of the plant growth substances have a number of analogs (for instance, more than 100 GAs have been identified) and there are frequent reports of other, as yet unidentified analogs.

The definition of a plant growth substance presupposes that it can be detected in the plant, and all of the above substances have been found in extracts or as volatiles from plant tissues. These analytical procedures usually involve organic and aqueous extraction of a plant or plant part, fractionation into acidic and neutral components, detection with a biological assay and by chemical methods (e.g., gas chromatography, high-pressure liquid chromatography), and finally identification and absolute proof of structure (e.g., mass spectrometry). For bioassay (short for *biological assay*), a test plant is selected that has a specific response to the substance in question. As an example, there are five dwarf corn mutants that are impaired in their GA synthesis and so normally do not grow tall. These will grow in specific response to GA applications. If application of a plant extract to one of these dwarf mutants causes it to grow, therefore, it is presumed that the extract contains GA. The *Avena* coleoptile curvature test is in a similar category; one of the oldest bioassays, it is nevertheless the best for identifying IAA and other auxins because their application causes a rapid increase in growth.

Auxins

At one time it appeared that the regulation of all plant growth was under the control of auxins, but it is now clear that differentiation, growth, and development are under the simultaneous and interacting control of all the plant growth substances. There is as yet no simple explanation for hormonal regulation of growth processes, but substantial new information has been developed through the use of molecular biology tools over the past decade. A gene has now been identified that encodes one cellular receptor for auxins (there may be several). Several of the steps in the pathway that communicate auxin presence to the rest of the cell have been described and the DNA elements that are crucial if auxins are to activate genes are known.

While the most important auxin is believed to be IAA, there are other auxins in plant tissues including 4-chloro-3-indoleacetic acid (4-Cl-IAA), phenylacetic acid (PAA), 3-indoleacetaldehyde (IAld), 3-indolepyruvic acid (IPA), 3-indoleacetonitrile (IAN), and the ethyl ester of indoleacetic acid (IAE). Even indolebutanoic acid (IBA), once thought to be a synthetic auxin and often used in promoting root development (see below), has been isolated from plants and may be a natural auxin. There is also a considerable body of literature about auxin derivatives and conjugates, some of which appear to be detoxification forms—indoleacetyl-aspartate, for example.

The paramount effect of auxin on growth is cell enlargement, although tissue culture studies show that auxin is also involved in cell division and root development from undifferentiated cells. The most significant and reproducible observation concerning growth response to auxin is that it is capable of "loosening" or increasing the plasticity of cell walls. As a result of this cell-wall loosening, internal turgor pressure against the cell wall causes cell expansion.

The mechanism by which IAA regulates growth is not completely understood; however, it is a spectacular feature of auxin that it acts both as a stimulator and an inhibitor of growth and that different plant parts—shoots, buds, and roots—show differential sensitivity to auxins. For example, the auxin-like herbicide 2,4-D (2,4-dichlorophenoxyacetic acid) stimulates cell enlargement at low concentrations, but at higher concentrations it causes inhibition. At still higher concentrations it is phytotoxic.

An unusual feature of the auxins is that when applied to plants at relatively high concentrations they stimulate the production of ethylene. Endogenous ethylene levels are thought to be regulated in part by the auxin level in the tissues, and this ethylene is likely involved in a number of growth processes. Certainly one effect of the herbicidal auxin 2,4-D is to promote large-scale production of ethylene, which may inhibit cell elongation while stimulating lateral cell enlargement. The result is a thickening of the stem. Auxin-induced ethylene may also inhibit root growth and cause downward bending of the leaves (epinasty). Thus, the stature of a plant is influenced dramatically by auxin synthesis and translocation.

Abscission, the shedding of organs—leaves, fruit, or flowers—is, in part, controlled by auxin. Apparently, auxin has two different and seemingly antagonistic roles in abscission. Applied auxin promotes ethylene production, which hastens abscission, probably by hastening senescence of the leaf. However, auxin is required for the viability of the cells in the abscission zone. Thus, abscission under natural conditions is partly under the control of a delicate balance of ethylene and auxin. Foliar application of auxin may delay abscission because it maintains cellular integrity, despite the presence of high levels of ethylene. It is probable that in addition to its well-documented effects on cell wall extension, auxin also participates biochemically in systems that maintain the function of plant cells.

Auxin is thought to be involved in many other plant processes, particularly fruit set and development and senescence.

Gibberellins

The gibberellins are a family of plant growth substances, the first of which were discovered in Japan in the fungus *Gibberella fujikuroi*, but which have subsequently been isolated from plants. These compounds share a similar structure whose configuration varies slightly in the different gibberellins. There are now more than 100 known GAs, identified as gibberellin A_1 (GA_1), gibberellin A_2, gibberellin A_3, etc. Not all of these are biologically active, and some show a marked specificity in particular processes. For example, GA_4 and GA_7 are about 5,000 times more active in stimulating elongation of cucumber hypocotyls than either GA_1 or GA_3, although the latter gibberellins are extremely active in stimulating elongation in other plants.

The most obvious effects of the gibberellins are cell elongation and division. These physiological responses attracted investigators in the mid-1950s, when a great deal of the horticultural and physiological work was initiated. The effect of gibberellins on plants is to promote cell division, enlargement, or both, particularly in the subapical meristem region, about 100 microns below the stem tip. Treatment of a radish plant, for example, will result in bolting and stem elongation roughly equivalent to that which occurs normally in response to a photoperiod or temperature treatment that will induce stem growth and flowering. It is now known that one of the first responses of plants to their "measurement" of the photoperiod is change in the synthesis of active gibberellins. It is the change in a plant's internal gibberellin content that then influences bolting and flower development.

There are a few instances in which GAs and auxins act in parallel ways. One example is in stimulating cucumber hypocotyl elongation. At present it appears that auxin is necessary for gibberellin action in elongation, and it appears possible that gibberellin acts through auxin in some processes. It is already known, however, that the GAs have certain effects as discussed below.

One of the developments that has led to an understanding of gibberellin action has been the discovery of a series of growth retardants, such as (2-chloroethyl)-trimethyl-ammonium chloride (chlormequat), that inhibit stem elongation. This inhibition is reversible by treatment with a GA. The retardants act as inhibitors of gibberellin biosynthesis, so the treatment results in a reduction in cell division and enlargement in the subapical region. It is possible to keep plants dwarf or short in stature by applying retardants to their soil or foliage. Additional insight into gibberellin action comes from work with protoxidone calcium (the active ingredient in Apogee), which acts within the plant to inhibit biosynthesis of gibberellin, resulting in less cell elongation and less vegetative shoot growth.

The gibberellins are involved in a variety of developmental processes that affect the yield and quality of crops. Dormancy, a condition in which a seed fails to germinate or a bud to elongate, is affected by plant growth substances. Buds of dormant potatoes can be stimulated to sprout by low levels of GA, and seeds of many species, including lettuce and celery, will germinate under unfavorable temperature conditions in response to GA. Gibberellins are not currently registered in California for use on lettuce and celery seed, but they are used for sprouting potatoes.

Gibberellins also appear to be involved in the process of fruit set and perhaps enlargement. They are capable of inducing parthenocarpy in more species than auxin, which was at one time considered the primary substance involved in regulating that process.

Whereas auxin has been shown to induce flowering in three or four species, GA stimulates flowering in many (but not all) species that otherwise require either low temperatures or long days. On the other hand, GAs have also been used to inhibit flowering in a number of species, particularly in perennials and short-day plants. It appears that the gibberellins are not, after all, the classically hypothesized flowering substance, called *florigen*.

Moreover, the nature of the flowering induced by GA is not typical. For example, in long-day plants, GA almost invariably induces stem elongation first and then flowering; the concentrations required to do so are usually relatively high. Gibberellins are undoubtedly involved in the flowering process, probably directly in relation to bolting and seedstalk elongation and indirectly in the sequence of events leading to flowering in responsive species.

In addition to their ability to enhance flower initiation, GAs influence sex expression in a number of plants. In a gynoecious (totally female) cucumber plant, GAs are capable of inducing and maintaining male flower formation. For example, application of GA_4 or GA_7 at the time of cucumber flower initiation (when the first true leaves are expanding) will cause male flower differentiation and such differentiation will be maintained over an extended period of time. Use of GA on cucumber is not currently registered in California.

Gibberellins can enhance the synthesis of plant enzymes. The embryo of a barley seed produces a GA that is translocated to the aleurone layer, and there activates the synthesis or release of α-amylase and other hydrolases involved in the breakdown of cell walls and storage materials (i.e., proteins and starch) in the endosperm. The breakdown products (amino acids and simple sugars) are transported to the embryo for its growth. Thus, early seedling growth is ultimately dependent on the availability of GA. Barley seeds whose embryos are partially damaged germinate very slowly, if at all. Gibberellin may therefore be thought of as a functional communicator between the embryo and the aleurone layer in the process of seed germination. This biological function of GAs is the basis for their use in promoting the malting of grains as part of the brewing process.

Cytokinins

The cytokinins are a group of plant growth substances that are important in regulating plant growth and development, especially cell division and apical dominance and senescence. Since cytokinins have many varied effects, they are difficult to define. One definition that has been suggested, however, is that they are compounds that, irrespective of other activities, promote cytokinesis (cell division) in cells of various plant organs.

Today, cytokinins or natural products that contain cytokinins (e.g., coconut milk, yeast extract) are a standard addition to nearly all tissue culture media. The most spectacular effect of cytokinins on development is the control of tissue and organ differentiation. Kinetin (one of the cytokinins) and IAA do a "juggling act" in determining the nature of differentiation in many plant tissue cultures. Thus a balanced level of kinetin and IAA results in production of only undifferentiated callus. An increase in the ratio of kinetin to IAA results in the formation of buds that can develop into shoots and even into entire plants. Conversely, an increase in the ratio of IAA to kinetin causes only roots to appear, and bud development is inhibited or prevented. As with all plant growth substances, the effect is not always universal and not all differentiation can be controlled by an auxin-cytokinin balance.

Germination of many seeds and buds is influenced by cytokinins. For example, although this method is not used commercially, kinetin can stimulate germination in lettuce seeds that fail to germinate at high temperatures (above 30°C [86°F]). It can also substitute for the red-light requirement for breaking seed dormancy at lower temperatures. It appears that one action of cytokinins in germination is to promote cotyledon expansion, which results in the rupture of the seed coat, permitting root and embryo elongation.

Cytokinins also participate in the control of budbreak. Apparently, cytokinins increase in concentration in axillary buds, promoting cell division and stimulating their expansion. Direct application of a cytokinin to axillary buds promotes growth and foliar application to some plants (such as petunias) results in a more highly branched plant as many axillary buds grow out. IAA is thought to inhibit the outgrowth, so the two substances play an antagonistic role in regulating apical dominance.

Another important role of the cytokinins is the regulation of leaf senescence. If a leaf is kept in the dark after treatment with cytokinin, it will retain chlorophyll and senesce more slowly than the untreated control leaf. If only half a leaf is painted with a cytokinin, the treated part will remain green while the other half turns yellow. Even a spot of cytokinin on a leaf will remain green while the rest of the leaf dies. In nature, unless a plant axis includes active roots, the leaves will not survive. It was found that cytokinins produced in the root are translocated to the leaf and there act to maintain leaf function, just as if the leaf had been treated with a cytokinin.

One implication of the evidence for cytokinin involvement in apical dominance and senescence is that it is involved in the mobilization of substances to the site of cytokinin production or treatment. When a cytokinin is applied as a droplet on a leaf, substances (as monitored using radioactive labeling techniques) move preferentially to the treated spot. As part of their senescence-delaying activity, cytokinins play a role in maintaining RNA and protein metabolism, thereby preserving the integrity of cells. They also inhibit the synthesis of proteases, which act to break down proteins; leaf protein loss is a clear indicator of leaf senescence.

Abscisic Acid

Abscisic acid (ABA) is a plant growth substance that was isolated by F. Addicott and his group at the University of

California, Davis, and subsequently by Wareing and Cornforth and associates in England. Addicott's compound, originally named Abscisin II and detected as a promoter of abscission, turned out to be the same as that isolated by Wareing and his associates in their search for dormin, a compound thought to be responsible for dormancy in buds. The compound was later renamed abscisic acid. Since the discovery of ABA, related compounds such as phaseic acid and xanthoxin have also been isolated. ABA is not used in agriculture at present, primarily because of its cost and the difficulty plants have taking in ABA that is applied to their surfaces. ABA has been shown to occur in numerous vascular plants and is probably present in all. It is found in amounts up to 14 milligrams per kilogram in soybean fruit.

ABA is central to the establishment, maintenance, and breaking of seed dormancy in many plants. ABA may not act alone in this process, and both cytokinins and gibberellins have been found experimentally to interact with ABA in regulating bud and seed dormancy. For example, direct application of GA_3 to potato buds promotes DNA and RNA synthesis and bud growth, whereas ABA prevents these developments.

As with dormancy, the role of ABA in abscission is not always evident. Abscisic acid applied in certain assay systems certainly does promote abscission. Abscisic acid increases in lupine flowers and cotton bolls at the onset of abscission. When ABA is applied to intact plants, however, it has virtually no ability to promote abscission except at very high, non-hormonal or physiological concentrations. This may be a reflection of the difficulty plants have in taking ABA in through their surface cuticle layer.

The effects of ABA on RNA and protein synthesis are also of interest because these processes are so basic to the process of senescence. When a leaf is treated with ABA, it turns yellow earlier than the control treated with water. There is evidence that cytokinin inhibits the senescence that is accelerated by ABA.

In maturing fruit, ABA levels increase dramatically as levels of cytokinins, auxins, and gibberellins become vanishingly low. This increase in ABA may be involved in enhancing the onset of fruit ripening and senescence and is of particular interest to researchers who study ripening control in fruit whose ripening is not promoted by ethylene (see below).

ABA, together with carbon dioxide (CO_2), plays an important role in controlling stomatal closure in leaves under stress conditions. Foliar applications of ABA result in closure within a few minutes, which may explain why ABA-treated plants are rendered tolerant to water stress. Endogenous ABA increases dramatically within stressed plants and decreases in level after water is provided. ABA appears to be involved in regulating other plant responses to water shortage and an assortment of other environmental stress factors (e.g., high and low temperatures and wounding).

Mutant plants have been identified that are known to be deficient in ABA. Their seedlings are unable to close their stomata during times of water stress. That is, they show a "wilty" phenotype, because water shortage inevitably causes wilting.

Ethylene

Ethylene (C_2H_4) is a highly volatile, relatively unreactive gas produced by all plants that has profound effects on virtually every phase of plant growth and development. Initially detected in illuminating gas, ethylene was shown to cause or enhance the degreening of citrus fruit—the final postharvest processing step when preparing oranges for market.

Ethylene has been the subject of study, primarily by postharvest physiologists. Today, ethylene is known to be involved in virtually every plant growth process. It is particularly interesting that, at one time, ethylene was considered only as a senescence or ripening factor. In some fruit, ethylene appears to be the main plant growth substance associated with the induction of ripening.

IAA and ethylene frequently exhibit similar effects. The reason is that high-concentration applications of auxin to plants stimulate the plants' ethylene production. Wounding has a similar effect. Although auxin stimulates ethylene production, not all ethylene effects can be attributed to auxin. At times the two substances act independently. Auxin has been shown to induce flowering in a very few species, including pineapple.

Ethylene, too, has the ability to induce flowering in pineapple and other bromeliads. Similarly, both auxin and ethylene promote the production of female flowers on monoecious cucumber plants. This property is useful in the production of hybrid cucumbers.

Despite the fact that ethylene generally acts as a plant growth inhibitor, the gas has a spectacular effect on the enlargement of fig fruit and, surprisingly, rice seedlings, especially the coleoptiles. Ethylene application midway during fig fruit development results in a remarkable increase in fruit expansion. Ethylene has also been shown to stimulate the enlargement of peach fruit, although this effect is not nearly so striking. Ethylene products are not currently used in commercial production of peaches, figs, or rice in California.

Ethylene may also play a role in dormancy, as it has been shown to stimulate germination in certain seeds highly sensitive to light. It is also known to affect sprouting in potatoes, however; short-term treatments stimulate sprouting, but long-term applications suppress it. It is possible that ethylene is acting as an endogenous inhibitor of sprouting during the rest period, but there is no evidence for this at present. Lettuce, clover, peanut, and other seeds have been stimulated to germinate by ethylene, although the mechanism by which ethylene acts in germination is unknown.

Ethylene enhances cellulase synthesis in the abscission zones of bean plants and it appears to be an important factor in abscission. The effect of ethylene on abscission is similar to that of auxin, which, at low concentrations, may itself stimulate ethylene production. At higher concentrations, ethylene and auxin seem to be antagonistic: when explants (excised plant parts) are treated with high levels of auxin they overcome the influence of ethylene. Since auxin can stimulate ethylene production, it appears that the abscission-stimulating effects of low levels of auxin are attributable to its promotion of the plant's ethylene production. High levels of auxin, however, act to maintain cellular integrity in the abscission zone. Thus the two substances can act either similarly or antagonistically, depending on their concentration.

PLANT GROWTH REGULATORS OF AGRICULTURAL INTEREST

The six classes of plant growth regulator closely parallel the five major classes of plant growth substance. The plant growth regulator classes are

- auxins
- gibberellins
- cytokinins
- growth retardants/inhibitors
- ethylene
- others

Only those compounds that are in current use or that are of potential use for specific agricultural applications are discussed in detail in this section. Consult your local agricultural authorities and read the pesticide labels to determine which chemicals are approved for commercial use. In general, plant growth regulators (as opposed to herbicides) may be considered as supplementary chemical tools that are potentially useful in plant or crop management. As such they are not absolutely essential, although growers depend on them increasingly for certain applications as described below. For information about their use on specific crops, refer to the appropriate section of this book.

Plant growth regulators are generally applied in small amounts that vary from a few grams to a pound per acre. Some are compatible with others PGRs and also with fungicides and other pesticides; however, interactions with other PGRs and other agricultural chemicals can diminish or negate their effectiveness. When possible, it is best to apply PGRs by themselves. Despite the relatively small quantities used, the potency of most PGRs is such that many of them are potentially dangerous to nontarget plants. For this reason, applications must be made under the strictest control to avoid drift. When mixing or calculating concentrations, a mistake in placement of a decimal point may spell the difference between success and disaster. You must also exercise caution to apply compounds only when wind drift can be avoided. Common sense and strict attention to the manufacturer's label directions are essential components of good application practice.

Auxins

Shortly after the spectacular effects of IAA became known, a variety of synthetic substances were found to have similar, and often greater, effects than the naturally occurring substances. Of greatest interest was the ability of high concentrations of auxin to morphologically alter or selectively kill plants. This early observation led to the discovery of 2,4-D, an herbicide that selectively kills broadleaf plants.

Another application of commercial importance is the use of indolebutanoic acid (IBA) (formerly known as indole-3-butyric acid) and 1-naphthaleneacetic acid (NAA) for initiation of rooting, especially in certain difficult-to-root stem cuttings. NAA and naphthaleneacetamide have been used on apples, pears, and olives to promote fruit thinning.

Two auxins, 4-chlorophenoxyacetic acid (4-CPA) and beta-naphthoxyacetic acid (BNOA), have been used to induce fruit set in tomatoes under low temperature conditions, although they are not currently registered for this use in California. The resulting, usually parthenocarpic (seedless) fruit may be larger than normal, but may also be "puffy"—that is, the locules may be devoid of jelly. Enlargement of grape berries is stimulated by 4-CPA, but this, too, is a practice not used in California. Developing orange fruit can be thinned using 2,4-D or NAA and, paradoxically, the same compound is useful in preventing abscission of mature oranges and other citrus. Triiodobenzoic acid (TIBA), sometimes an auxin antagonist, promotes axillary branching and actually changes the shape of soybean plants. Desirable wide branch angles in apple trees are also favored by TIBA applications.

Gibberellins

The gibberellins in greatest use today are GA_3 and a mixture of GA_4 and GA_7. While various effects of these compounds are known, by far the greatest use for any GA is on 'Thompson Seedless' grapes. Various prebloom and postbloom treatments have been recommended to achieve different effects on different varieties. Essentially all 'Thompson Seedless' grapes grown in California are treated to thin clusters and induce berry enlargement. Berry size of seedless 'Black Corinth' and one-seeded 'Emperor' grapes is enhanced by GA_3 as well. In 'Tinta Madeira' and 'Palomino' wine grapes, application of GA_3 promotes enlargement of the rachis, making it unnecessary to prune the cluster. Failure to thus "loosen the bunch" frequently results in berry rupture and subsequent infection. It should be noted that not all grape varieties respond to GAs with increased berry size.

Promotion of seedstalk elongation and flowering in many annual and biennial crops is a common response to the gibberellins, particularly GA_3, GA_4, and GA_7. In head lettuce, treatments applied to young plants facilitate emergence of the seedstalk, which otherwise would be trapped within the enfolding leaves. The method is used by some seed companies; however, it is essential to rogue off-type plants throughout the growing period, as GA-treated plants do not produce a typical head.

Preharvest sprays of GA_3 inhibit development of rind defects on navel oranges. In lemons, preharvest application of GA_3 retards degreening during storage. The result is extended storage life for the fruit and greater control of its availability on the market.

Cytokinins

Although not registered for use in California, the compound, benzyladenine (BA) is useful in inhibiting elongation and promoting axillary branching in certain flower crops, on holly, and on cherry, apple, and Christmas trees. Along with the stimulation of axillary branching, BA applications may also result in an intensification of flowering. Gibberellins are often combined with other cytokinins to promote branching and elongation.

Ethylene and Ethephon

Ethylene, a naturally occurring gas, promotes ripening in mature fruit. In bananas and market tomatoes, the gas is used to effect ripening at destination sites. Mature green fruit are shipped at low (12°C [53°F]) temperatures to slow ripening. On arrival at the terminal market, they are placed in ripening rooms where they receive an ethylene treatment that induces ripening.

Ethephon, a generic name for the synthetically produced ethylene-releasing compound (2-chloroethyl) phosphonic acid, is currently used for the promotion of fruit ripening in mechanically harvested canning tomatoes. Field applications are made early in the season when fewer than 20 percent of the fruit are pink. Later in the year when temperatures are lower, the compound is applied when fewer than 30 percent of the fruit show pink color. The ethylene released on contact with the plant promotes earlier ripening of the entire population of fruit that are mature enough to respond to the compound.

Initiation of female flower formation in cucumber is under the control of endogenous ethylene. The application of ethephon to pickling cucumbers when the second true leaf is expanded (coincident with the time of flower initiation) promotes female flower initiation. The result is that all flowers at about the first 5 to 15 nodes are female (depending on growing conditions), and are then available for pollination, whereas in untreated plants the first flowers are always male. As a result of ethephon treatment, the flowers at the lowest nodes set almost simultaneously, and therefore the fruit on all of the plants grow to a desirable size at approximately the same rate. For once-over mechanical harvest, you can obtain a good yield of reasonably uniform fruit concentrated near the center of the plant.

Pineapple and ornamental bromeliads are induced to flower by ethephon. The response is conditioned by the age of the plant, its variety, and environmental and cultural conditions.

The development of mechanical harvesting has hastened the search for substances that can promote abscission (shedding) of fruit. In states other than California, this application has been particularly useful for loosening cherries with foliage sprays of ethephon when the cherries are sufficiently mature for processing. Because ethylene promotes abscission of leaves as well as fruit, special attention must be paid to preparation of the spray solution and conditions at the time of application. The loosening of walnuts for mechanical harvesting and separation of the nut meats from the shell is markedly enhanced by ethylene as well as ethephon.

Ethylene synthesis in the plant can be inhibited or its action blocked by aminoethoxyvinyl glycine (AVG), sold as ReTain, or with silver ion (as silver thiosulfate). Recently, the volatile gas methylcyclopropene (MCP, sold as EthylBloc) has been found to competitively inhibit ethylene action. These products are receiving a great deal of commercial interest in postharvest handling areas.

GROWTH INHIBITORS AND RETARDANTS

A number of compounds have the unique property of inhibiting elongation and other processes while doing minimal (tolerable) or no damage to the treated plant. These have proved useful for producing small potted plants, inhibiting excessive elongation in a variety of shrubs and trees, and altering plant conformation.

The oldest continually used plant growth inhibitor is maleic hydrazide (MH), which is effective in reducing sprouting on onions and potatoes and thereby extending storage life. Ideally, the onion bulbs are sprayed when they are fully mature, with five to eight leaves that are on the verge of falling over. For inhibition of potato sprouting, MH is sprayed on the foliage 4 weeks before harvest when the tubers are about 1 inch in diameter.

Maleic hydrazide is remarkably useful in inhibiting growth of trees, a special advantage in situations where they grow under power lines. Such chemical pruning has saved virtually all of the cost of hand pruning. The compound actually damages or kills the growing points of the trees, thus preventing further elongation. The toxic effect is barely noticeable and only for a short time after treatment, after which time it disappears.

Growth of grasses along ditch banks and hard-to-mow areas is controlled by spray applications of MH. The herbicide 2,4-D is compatible with MH, so it can be included as a combination spray to kill broadleaf plants.

Isopropyl-N-(3-chlorophenyl) carbamate (Chloro IPC) is an herbicide and a very powerful inhibitor of sprouting in stored potatoes. The compound is applied as a gas in storage, as a dust on the tubers, or on strips of paper dispersed through a pile of potatoes. Internal sprouting is an undesirable side effect of Chloro IPC treatment of potatoes. In such cases, the sprouts begin to elongate as temperatures increase in common storage. The direction of growth is inward so that the defect often goes unobserved in the potato until after it is sold.

Succinic acid (2,2-dimethyl) hydrazide (SADH) has helped revolutionize the ornamental plant industry because it retards elongation of stems of chrysanthemums, azaleas, hydrangeas, poinsettias, gardenias, and a variety of annual bedding plants.

Another growth retardant, chlormequat (2 [chloroethyl] trimethylammonium chloride, or CCC, or Cycocel), has proved to be a highly effective compound in reducing stem elongation. In Europe, tall straw varieties of wheat are sprayed with chlormequat to inhibit elongation and thereby prevent lodging. This use is not registered in the United States, but chlormequat is used in the floriculture industry where it has helped change potted-plant production methods. Poinsettia plants, for example, are dwarfed by soil application of the compound and can be sold as short-stemmed, branched, flowering plants.

Most growth retardants used today act by interfering with endogenous gibberellin synthesis or action. Ancymidol (α-cyclopropyl-α-(4-methoxyphenyl)-5-pyrimidinemethanol, or A-REST) is used to control stem elongation in poinsettias, Easter lilies, and tulips when applied as a soil drench or foliar spray and in bedding plants as a foliar spray. Paclobutrazol (1-(4-chlorophenyl)-4,4-dimethyl-2-(1,2,4-triazol-1-yl)pentanol, or Bonzi) is used on bedding plants to control height and on seed grasses to reduce lodging. Daminozide (N-Dimethylamino) succinamic acid (B-Nine or Alar) is applied to bedding plants, poinsettias, azaleas, hydrangeas, and gardenias to reduce height. Uniconazole (1-(4-chlorophenyl)-4,4-dimethyl-2-(1,2,4-triazol-1-yl)-penteneol (Sumagic) is applied as a soil drench to control the height of dahlias, lilies, poinsettias, and chrysanthemums.

SAFE AND SUCCESSFUL USE

Plant growth regulators are considered pesticides according to state and federal law. Plant growth regulators therefore must be registered by the federal and state Environmental Protection Agencies (USEPA and CalEPA), their use reported to the Department of Pesticide Regulation (DPR), and all label instructions adhered to for their storage, transportation, and application.

Most of the registered growth regulators have relatively low acute and chronic toxicities in laboratory animals, but the full range of tests on nontarget organisms, movement, and persistence has not been completed for many potentially useful compounds. Those who recommend application should be familiar with registration requirements and limitations.

Several plant regulators occur naturally in the edible portions of plants. Those synthetic regulators used commercially also may be considered "safe" when employed in accordance with regulations established by the Environmental Protection Agency and the State of California.

Examples of hazards associated with some plant growth regulator products include corrosive effect (ethephon), flammability (ethephon, gibberellins), eye injury (ethephon, gibberellins), and bee hazards (carbaryl). All plant growth regulators can cause injury to plants if improperly used. Always check labels to identify hazards associated with each particular product.

Care should be exercised in applying growth regulators to avoid drift onto people, livestock, and other plant materials and to observe the waiting interval between application and harvest if one is indicated on the label. Because of their remarkable effectiveness, only a small amount of a regulator is usually needed. However, the temptation to apply "more for insurance," as is common with commercial fertilizers, can be disastrous for the treated plants and possibly for neighboring crops. A mistake in placement of a decimal point when diluting for application can result in a ruined crop. Plant growth regulators are generally applied in the parts per million (ppm) range. One ppm is equal to 1 mg per liter of solution. **FOLLOW INSTRUCTIONS ON THE LABEL.**

Mode of application is important when you are employing plant growth regulators. In some cases the same compound may be applied either as a foliar spray, as a soil drench, or by bark injection. Follow the label instructions carefully. The commercial use of most growth regulators has developed using ground-rig application involving solution amounts ranging up to 200 gallons per acre. Because most aerial application rates do not exceed 15 gallons per acre, you should conduct tests locally to determine whether aerial applications can yield results that are comparable to those for ground application. Both the chemical applicator and the crop owner are morally and legally responsible for preventing the aerial drift of chemicals onto the property of others.

Adjuvants or surfactants and pH adjustment are often recommended for inclusion in plant growth regulator solutions as a way to overcome water chemistry and surface tension problems and facilitate the spread and penetration of the chemical into the plant. These additional chemicals should only be used as recommended by the plant growth regulator's manufacturer, as they may influence the activity of the regulating compound. In general, regulators are reasonably stable in water. At least one of them, though, ethephon, must be used shortly after dilution since its volatile active component (ethylene) is released above a pH of 4.0. The commercial preparation is prepared at pH 2.0 in order to prevent release of the ethylene. Addition of the concentrate to water, which is usually near pH 7.0, results in release of the ethylene. The concentrate must be handled with care; it is acidic and can burn the skin.

Plant growth regulators, like all chemicals, vary in their activity in relation to formulation, concentration, environmental conditions (e.g., light intensity, humidity, temperature), target species, and time of application. A compound useful on one crop is not always useful on another crop, even if the harvested product (e.g., a fruit) is the same. Almost invariably, timing of application is a critical factor in obtaining a desired effect. Indeed, the stage of a plant's development is often the most critical factor in determining the final effectiveness of the applied plant growth regulator. Application of a PGR at the wrong stage of development may cause more harm than good. Correct knowledge about the relationship between the frequency of application to a plant, the PGR's persistence, its phytotoxicity, and the intended effect of the application is of paramount importance. This information normally will be carried on the label or in instructions usually available from the manufacturer. With some compounds, you may need to apply different amounts in different locations and at different times during the growing season in order to achieve your purpose. Proper and accurate calibration of application equipment is necessary to ensure that you use the correct amount of active ingredient. Regulators are often compatible with other compounds, such as herbicides and pesticides, but they should not be mixed with other compounds unless you are certain of their compatibility. Otherwise you may risk loss of the desired effect or damage to the crop. **FOLLOW INSTRUCTIONS ON THE LABEL.**

Selected Literature

Addicott, F. T. 1970. Plant hormone and the control of abscission. Biol. Rev. 45:485–524.

Addicott, F. T., and J. L. Lyon. 1969. Physiology of abscisic acid and related substances. Ann. Rev. Plant Physiol. 20:139–164.

Aldus, L. J. 1959. Plant growth substances. London: Leonard Hill Ltd.

Anonymous. 1961. Plant growth regulation. Ames: Iowa State University Press.

Anonymous. 1968. Plant growth regulators. Soc. Chem. Ind., Monograph no. 31.

Basra, A. S. (ed.). 2000. Plant growth regulators in agriculture and horticulture. Binghamptom, NY: The Haworth Press.

Creelman, R. A., and Mullet, J. E. 1997. Oligosaccharins, brassinolides, and jasmonates: Nontraditional regulators of plant growth, development, and gene expression. The Plant Cell 9: 1211–1223.

Davies, P. J. (ed.). 1995. Plant hormones: Physiology, biochemistry, and molecular biology. Dordrecht: Kluwer Academic Publishers.

Frederick, J. F. (ed.). 1967. Plant growth regulators. Ann. N. Y. Acad. Sci. 144.

Goldsmith, M. H. M. 1977. The polar transport of auxin. Ann. Rev. Plant Physiol. 28:439–478.

Hoad, G. V. (ed.). 1987. Hormone action in plant development: A critical appraisal. London: Butterworths.

Hooykaas, P. P. J. J., M. A. Hall, and K. R. Libbenga. 1999. Biochemistry and molecular biology of plant hormones. Amsterdam: Elsevier Science.

Kende, H., and J. A. D. Zeevaart. 1997. The five "classical" plant hormones. Plant Cell 9:1197–1210.

Klee, H., and M. Estelle. 1991. Molecular genetic approaches to plant hormone biology. Ann. Rev. Plant Physiol. Plant. Mol. Biol. 42:529–551.

Lang, A. 1970. Gibberellins: Structure and metabolism. Ann. Rev. Plant Physiol. 21:537–570.

Letham, D. S. 1967. Chemistry and physiology of kinetin-like compounds. Ann. Rev. Plant Physiol. 18:349–364.

Moore, T. C. 1989. Biochemistry and physiology of plant hormones. Second Edition. New York: Springer-Verlag.

Nickell, L. G. 1982. Plant growth regulators: Agricultural uses. Berlin, Heidelberg, New York: Springer-Verlag.

Pharis, R. P., and R. W. King. 1985. Gibberellins and reproductive development in seed plants. Ann. Rev. Plant Physiol. 36:517–568.

Phillips, I. D. J. 1971. The biochemistry and physiology of plant growth hormones. New York: McGraw Hill Book Co.

Pratt, H. K., and J. D. Goeschl. 1969. Physiological roles of ethylene in plants. Ann. Rev. Plant Physiol. 20:541–584.

Rappaport, L. 1970. Chemical regulation of plant development. Amer. Assoc. Advancement Sci. 147–180.

Reid, J. B., and S. H. Howell. 1995. The function of hormones in plant growth and development. In Plant hormones: Physiology, biochemistry, and molecular biology, P. J. Davies, ed., pp. 448–485. Dordrecht: Kluwer Press.

Schneider, G. 1970. Morphactins: Physiology and performance. Ann. Rev. Plant Physiol. 21:499–536.

Skoog, F., and D. J. Armstrong. 1970. Cytokinines. Ann. Rev. Plant Physiol. 21:359–384.

Taiz, L., and E. Zeiger. 1998. Plant physiology, Second edition. Sunderland, MA: Sinauer Associates. (An excellent text. Chapters 19–23 are most relevant.)

Vardar, Y. (ed.). 1968. Transport of plant hormones. New York: American Elsevier.

Weaver, R. J. 1972. Plant growth substances in agriculture. San Francisco: W. H. Freeman.

Wittwer, S. H. 1971. Growth regulants in agriculture. Outlook on Agriculture. 6(5):205–17.

Reference Tables 2 through 6

Mode of Action and Common Uses of Plant Growth Regulators in California Agriculture

NOTE: The following tables give common examples of plant growth regulators and their uses in California agriculture. This is not, however, an exhaustive list. Other plant growth regulators are available, and most PGRs mentioned here have uses not listed in the tables. Although these uses are registered in California at the time of this writing (July 2002), registrations do change frequently, and some materials may no longer be available by the time you read this. Always check labels and check with local authorities for currently registered uses.

The Crop Data Management Systems Web site (http://www.cdms.net) is a good source for current label information.

Table 2. Mode of action and uses of auxins.

Auxins cause cell enlargement by loosening cell walls followed by cell expansion, stimulate production of ethylene, promote or delay abscission, and initiate adventitious roots.

Naturally occurring (endogenous) auxin: 3-indoleacetic acid (IAA)

Compound	Common name	Examples of uses for California
2,4-dichlorophenoxyacetic acid	2,4-D	Citrus: delays and reduces abscission and fruit drop; improves fruit size; increases storage vitality of lemons. Herbicidal use: selectively kills broadleaf plants.
1-naphthaleneacetic acid	NAA	Apples: prevents preharvest drop; increases fruit thinning. Citrus: promotes abscission; increases fruit thinning; aids in rooting cuttings. Olives: increases fruit thinning; increases fruit size and promotes annual bearing. Ornamentals: inhibits basal sprouting of woody ornamentals; promotes rooting; prevents nuisance fruit from forming. Pears: prevents preharvest drop. Other: aids in rooting cuttings of fruits and nuts.
Ethyl 1-naphthalene acetate	ethyl ester of NAA	Apples, citrus (non-bearing), nectarines, olives, pears, woody ornamentals: controls sprouting and sucker growth.
Indole-3-butyric acid (Indolebutanoic acid)	IBA	Nurseries: aids in rooting cuttings of fruits and nuts, including citrus.
3-indoleacetic acid	IAA	Nurseries: helps with rooting cuttings of citrus.
Other auxins		
4-chloro-3-indoleacetic acid	4-Cl-IAA	
Phenylacetic acid	PAA	
3-indoleacetaldehyde	IAld	
3-indolepyruvic acid	IPA	
3-indoleacetonitrile	IAN	
Ethyl ester of indoleacetic acid	IAE	
4-chlorophenoxyacetic acid	4-CPA	
Beta-naphthoxyacetic acid	BNOA	

Table 3. Mode of action and uses of gibberellins.

Gibberellins stimulate cell division, elongation, and enlargement (especially at the subapical meristem), improve fruit set and enlargement, induce parthenocarpy, and play a role in seed and bud dormancy.

Naturally occurring (endogenous) gibberellins: There are many — more than 100.

Compound	Common name	Examples of uses for California
Gibberellin A_{4-7} (GA_{4-7})	Various brand names, including some formulated with cytokinins	Apples: increases thinning, which improves fruit size; promotes lateral budbreak, which increases branching on young apple trees; improves apple shape; reduces russeting of apples. Cherries: promotes lateral budbreak, which increases branching on nonbearing cherry trees.
Gibberellin A_3	Gibberellic acid GA_3	Artichokes: increases earliness and uniformity of bud development (advances harvest) of artichokes. Celery: increases plant height and early yield of celery. Sweet cherries: delays maturity, increases fruit firmness, size, and quality. Citrus: delays pre- and postharvest fruit maturity, senescence, rind decay, rind coloring, and over-ripening; increases fruit set. Grapes: reduces fruit set of seedless table grapes, which increases fruit size; elongates clusters of seedless table grapes; increases fruit size of seedless and seeded table grapes; reduces fruit set and cluster compaction of raisin grapes ('Thompson Seedless') and improves airstream grade by increasing fruit weight and length; increases berry size of midget raisins (cultivar 'Black Corinth'); reduces fruit set and cluster compaction of wine grapes, which reduces bunch rot, but may reduce fruitfulness the following year. Lettuce: induces uniform bolting and increases seed production of lettuce. Potatoes: stimulates uniform sprouting and breaks dormancy of seed potatoes. Rhubarb: reduces the chilling requirement for forcing rhubarb.
Other gibberellins		Rice: improves seedling emergence. Stone fruits: improves fruit firmness, delays harvest, and thins fruit. Sweet cherries: delays maturity, increases fruit firmness, size, and quality.

Table 4. Mode of action and uses of cytokinins.

Cytokinins promote cytokinesis (cell division), play a role in tissue differentiation, budbreak, and apical dominance, delay senescence, and promote cotyledon expansion.

Naturally occurring (endogenous) cytokinins: zeatin, zeatin riboside, and others.

The cytokinin kinetin occurs in herring sperm. Although historically used to break seed dormancy, kinetin is not currently used commercially in California.

Compound	Common name	Examples of uses for California
N-(phenylmethyl)-1H-purine-6-amine	Several formulations, made with GA_{4-7}	Apples: increases thinning, which improves fruit size; promotes lateral budbreak, which increases branching on young apple trees; improves apple shape; reduces russeting of apples.
N-(2-chloro-4-pyridyl)-N'-phenylurea	Forchlorfenuron CPPU	Grapes: increases fruit set and berry size; delays fruit maturation; increases size and firmness of seedless table grapes when applied with GA_3 (experimental use only).

Table 5. Mode of action and uses of ethylene.

Ethylene is involved in many plant processes including ripening of some fruits, helping break dormancy, and promoting abscission.

Ethylene is a gaseous hormone produced naturally in many plant tissues.

Compound	Common name	Examples of uses for California
Ethylene	ethylene	Citrus: degreens harvested citrus.
(2-chloroethyl) phosphonic acid	ethephon	Apples: increases flower bud formation, fruit thinning, and fruit loosening; promotes uniform ripening reddening of apples. Barley: reduces lodging by stiffening stem and reducing plant height. Cotton: increases rate of maturation and opening of bolls. Curcurbits: hastens maturity, abscission of cantaloupes; on cucumbers and squash increases the number of female flowers and suppresses male flowers (useful in creating hybrid seed); increases fruit yield for pickling cucumbers. Grapes: enhances pigment accumulation in table grape skins (red and black cultivars); increases the amount of table grapes that can be harvested in first picking, but softens fruit; advances maturity of raisin grapes ('Thompson Seedless,' 'Fiesta'). Ornamental, nursery plants: eliminates nuisance fruit formation on ornamental trees; stimulates lateral branching of azaleas, geraniums; controls plant height of daffodils and hyacinths; induces flowering of some bromeliads; promotes earlier defoliation of roses. Peppers: accelerates maturation, uniformity, and fruit coloring of peppers. Pineapple, bromeliads: induces flowering. Tomatoes: advances maturation and color uniformity of both fresh and processing tomatoes, pre- and postharvest. Walnuts: advances hull splitting and fruit abscission of walnuts. Wheat: reduces lodging by stiffening stem and reducing plant height.

Table 6. Mode of action and uses of growth inhibitors and retardants.

This group of plant growth regulators contains a variety of unrelated substances, many of which can be classified as gibberellin biosynthesis inhibitors, terminal bud inhibitors, or ethylene synthesis inhibitors.

Note: One compound that is often included in this group is the plant hormone abscisic acid (ABA). ABA acts by promoting abscission and regulating dormancy and stomate closure. ABA is not used in agriculture as it is too costly and difficult to apply.

Group and common names	Mode of action	Examples of uses for California
Carbamates Carbaryl Chloro IPC		Apples: carbaryl insecticide acts as a fruit thinner. Potatoes: chloro IPC prevents or delays postharvest sprouting.
Gibberellin biosynthesis inhibitors Ancymidol Chlormequat Flurprimidol Mepiquat chloride Paclobutrazol Prohexadione calcium Uniconazole	All of these substances inhibit the synthesis of gibberellin in plants, resulting in reduced cell elongation and internode compression, and therefore, shorter plants.	Apples: prohexadione calcium reduces shoot growth and fireblight. Cotton: mepiquat chloride increases boll retention, reduces vegetative growth, and stimulates earlier maturity. Greenhouse & nursery plants: ancymidol, chlormequat, and paclobutrozol. Ornamentals: flurprimidol, uniconazole, and paclobutrazol on trees. Wheat: chlormequat reduces stem elongation (not registered in U.S.).
Terminal bud inhibitors Dikegulac-sodium Mefluidide Maleic hydrazide, MH	Inhibits or kills meristematic tissue, which reduces terminal growth and may stimulate lateral buds resulting in short-stature plants. Maleic hydrazide inhibits cell division.	Onions: maleic hydrazide prevents or delays postharvest sprouting. Ornamentals: dikegulac-sodium reduces or eliminates bloom and fruit set on landscape trees; mefluidide reduces growth of woody ornamentals; maleic hydrazide retards growth of ornamental trees and shrubs. Potatoes: maleic hydrazide prevents or delays postharvest sprouting. Turfgrass: mefluidide reduces turf growth and seed production; maleic hydrazide reduces growth.
Ethylene synthesis inhibitors Aminoethoxyvinyl glycine, AVG	Postharvest uses to reduce the rate of ripening.	Apples: delays maturity of fruit to allow for later harvest; reduces preharvest fruit drop; increases fruit color and size.
Hydrogen cyanamide (H_2CN_2)	Inhibits the enzyme catalase.	Grapes: Increases growth uniformity of early season table grapes by increasing uniformity of budbreak.
SADH Succinic acid Daminozide		Ornamental plants: retards elongation of flower stems (hydrangeas, chrysanthemums, azaleas, poinsettias, gardenias) and bedding plants; promotes flowering and reduces plant height of greenhouse and nursery plants.

2

Plant Growth Regulators for Fruit and Nut Crops

STEPHEN M. SOUTHWICK, Pomologist,
Department of Pomology, University of California, Davis

For more than 50 years, plant growth regulators (PGRs) have been used successfully to thin apple fruit at bloom time (dinitros) and prevent preharvest drop of apples and pears (NAA), and to help root cuttings (NAA and IBA). Ethylene in the form of ethephon or ethrel has been used to hasten nut and fruit harvest and to hasten the development of red color in apples. Growth regulators that reduce shoot length on fruit trees are currently under development. Many PGRs initially appear to be useful in production agriculture, but inconsistencies in performance, side effects, and risks involved in their use may outweigh any advantages. As a result, very few PGRs are commonly used in fruit production. All the same, at this writing more PGRs than ever before are available for deciduous fruit production.

Any recommendations that you make for using PGRs should be carefully backed by extensive testing on many cultivars in several districts and under a range of environmental and cultural conditions. The rapid rate of change in cultivars, however, combined with grower interest in maximizing their opportunities for profit, mean that growers often have to make application decisions based on imperfect information. Problems that commonly result from PGR use include non-performance and over-performance. Such variations in performance from year to year and cultivar to cultivar are fairly common with PGR use. Variations in performance for the most part result from weather conditions and spray distribution that may affect PGR uptake. Many PGRs penetrate the leaf and are translocated to fruit or other plant parts, so the amount and rate of absorption can be very important to predicting PGR performance. Variations in performance may also result when the applicator uses an improper pH in the spray mix or mismanages the spray application (for example, too little spray volume or application before rain). Surfactants and adjuvants may improve consistency in PGR performance. Some PGRs can also be phytotoxic, but this is rare if applications are kept within the label recommendation. As a rule, PGRs perform less consistently than herbicides, insecticides, or fungicides, but even so, they can be very useful in commercial fruit production. To be of practical commercial use, a PGR must perform with a degree of consistency and be safe to use.

Production and environmental factors can lead to variations in PGR performance, so you need to take careful note of all conditions existing when the chemicals are applied, and the spray application must be precisely timed. After application, take note of any adverse side effects on tree growth and fruiting during current and future years.

Plant growth regulators should not be applied with pesticides, oils, or surfactants until their performance both alone and in such combinations is well known. In most cases, wetting agents, adjuvants, oils, and emulsifying agents substantially increase the absorption and activity of plant growth regulators. Only healthy, normal trees should be treated, as weak trees are usually more likely to be injured. *Carefully read current labels for all materials, and note cautions on combinations, timing, cultivars, grazing livestock in sprayed orchards, and use of fruit pomace.*

FRUIT AND ORNAMENTAL WOODY PLANTS

General

NAA — 1-Naphthalenacetic acid

IBA — Indole-3-butyric acid

Dormant wood cuttings or soft herbaceous cuttings will in many cases root more completely if treated with NAA or IBA. Each plant species has its own optimal concentration, but general guidelines are as follows.

Dilute-soak method. The basal 1 inch of woody cuttings is soaked approximately 24 hours in solutions of 20 to 200 ppm of a root-promoting substance.

Quick-dip method. Basal ends of herbaceous and woody cuttings are dipped in solutions of 500 to 10,000 ppm IBA for about 5 seconds.

Dry powder method. Commercially prepared dry powder rooting preparations may be available. Dip the basal end of each cutting in the powder. Follow the directions on the package.

Reference

Hartmann, H. T., D. E. Kester, F. T. Davies, Jr., and R. L. Geneve. 1997. Plant propagation: Principles and practices. Sixth edition. Upper Saddle River, NJ: Prentice-Hall.

Dormex — hydrogen cyanamide

Dormex is a plant growth regulator used to stimulate budbreak on apples, sweet cherries, kiwifruit, and table grapes in California. In addition, Dormex is being used on apricot, plum, peach, blueberry, and pear elsewhere in the world and has been tested on many other fruit crops. Typically, growers use Dormex in years where the chilling has been less than desirable for a crop species or cultivar. Dormex is also used to promote earlier budbreak, which can lead to earlier-ripening fruit. Dormex is a restricted-use pesticide. Application personnel must not consume alcoholic beverages prior to, during, or within 24 hours after handling the product (check label for specific warnings).

Apples

Ethrel brand of ethephon — (2-chloroethyl) phosphonic acid

Ethephon may reduce shoot growth and promote increased flower bud formation on young trees in the season after application. In the spring, ethephon is used to thin small fruit. Later in the season, ethephon is also used to loosen fruit and promote uniform ripening and red fruit color of red-fruited cultivars. Use of this material to induce flower bud formation and loosen fruit has been very limited in California.

To promote more uniform ripening and red fruit color in 'Red Delicious' and some other red cultivars, ethephon must be applied 2 to 3 weeks before normal harvest and 1 to 2 weeks before desired harvest. Use enough water for uniform spray coverage. If treated trees are to be harvested at their normal maturity date, a preharvest drop-control chemical needs to be applied to prevent excessive fruit drop. With regard to fruit thinning, most applications are made shortly (i.e., 10 to 20 days) after full bloom. Ethephon may also be mixed with other thinning agents like Amid-Thin brand of NAD, or SEVIN brand of carbaryl. Amid-Thin has been found to produce smallish (pygmy) fruit in 'Delicious' and as a result is not recommended on that cultivar and should be evaluated for that problem before use on other cultivars. Chemical thinning programs vary by cultivar, region, and season, but are necessary to promote regular bearing and adequate fruit size.

Reference

Micke, W. C., R. H. Tyler, and J. T. Yeager. 1977. Growth regulators affect apple maturity. Calif. Agric. 31(3):15.

Fruitone N, Liqui-Stik brands of NAA — 1-naphthaleneacetic acid, sodium salt; 1-naphthaleneacetic acid, ammonium salt, respectively

Fruit thinning. For thinning, NAA is sprayed at petal fall or up to 3 weeks after full bloom, depending upon the cultivar. Generally, temperatures should be between 70° and 75°F (21° and 24°C) when NAA is applied. Applications should be avoided when temperatures fall below 60°F (16°C) or above 80°F (26°C). Concentrations range from 2.0 to 20 ppm, depending upon cultivar, NAA formulation, tree vigor, previous cropping, thinning requirements, and other factors. California growers use it on the 'Golden Delicious' and 'Red Delicious' cultivars in the Sierra Nevada foothills. NAA sometimes causes over thinning, little thinning, some flagging of foliage, stunting of shoot growth, and reductions in fruit size. Fruit size reductions have been noticed, especially with the amide form of NAA when applications are made after petal fall. When combined with certain surfactants or wetting agents, the quantity of NAA can be reduced, or the thinning effect may be increased.

Preharvest fruit drop control. NAA is also used to control preharvest drop. Spray may be effective for drop control within 1 to 2 days after application. Cool weather may reduce the onset of effectiveness. Typically applications are made when the first sound, uninjured fruit begin to drop or 7 to 14 days before harvest, depending upon formulation. Aerial application may be possible. Although fruit drop can be controlled, it has been shown that NAA may also enhance the fruit ripening of various apple cultivars.

References

Flint, M. L. (tech. ed.). 1999. Integrated pest management for apples and pears. Second edition. Oakland: University of California Division of Agriculture and Natural Resources, Publication 3340.

Koch, E. C., et al. 1992. Commercial apple growing in California. Oakland: University of California Division of Agriculture and Natural Resources, Leaflet 2456.

Washington State University. 1990. Spray guide for tree fruits in eastern Washington. Washington State University Extension Bulletin 419.

Amid-Thin (Amide) brand of NAAmide or NAD — naphthaleneacetamide

Amid-Thin is used as a chemical thinner for apples. A single application is made at petal fall. Do not use on 'Red Delicious,' as deformed, "pigmy" fruit may develop. Carbaryl, oils, surfactants, and wetting agents can increase the thinning obtained from Amide, so mixing with other chemicals may be dangerous.

Sprout or "sucker" control. Tre-Hold brand of Ethyl 1-naphthalene acetate, ethyl ester of NAA, is closely related to NAA and is used in apples, pears, olives, non-bearing citrus, non-bearing nectarines, and woody ornamentals to control sprouting and sucker growth. It is often applied in a formulated spray in concentrations that are much higher than for other reported uses, approximately 1.15 percent v/v active ingredient. Tre-Hold typically is applied topically to severe pruning wounds on limbs or along limb sections that are prone to suckering or near trunks where suckering may be excessive. The application must be made before sprouting or suckering occurs (i.e., early spring). When using this product, take care to avoid drift, since contact with other nontarget fruit may lead to reduction in its growth.

References

Flint, M. L. (tech. ed.). 1999. Integrated pest management for apples and pears. Second edition. Oakland: University of California Division of Agriculture and Natural Resources, Publication 3340.

Koch, E. C., et al. 1992. Commercial apple growing in California. Oakland: University of California Division of Agriculture and Natural Resources, Leaflet 2456.

Washington State University. 1990. Spray guide for tree fruits in eastern Washington. Washington State University, Extension Bulletin 419.

SEVIN brand of carbaryl — 1-naphthyl N-methyl carbamate

This insecticide is an effective fruit thinner in many areas of the world. Applications are made from 80 percent (early) petal fall to 16 mm fruit size, usually when the largest fruit is between 10 and 16 mm (10 to 25 days after full bloom), depending upon formulation. Spray applications may be made twice during the effective period. Doses vary with cultivar, growing location, tree age, production practices, and other factors. Lower doses are recommended on easily thinned cultivars. Some formulations of Sevin are highly toxic to honey bees, mite predators, and rust mites. The Sevin brand XLR Plus carbaryl insecticide is preferred if you want to avoid honey bee injury. To minimize damage to predaceous mites, Sevin sprays are directed towards tops of trees; trunks and lower limbs should not be sprayed.

References

Koch, E. C., et al. 1992. Commercial apple growing in California. Oakland: University of California Division of Agriculture and Natural Resources, Leaflet 2456.

Washington State University. 1990. Spray guide for tree fruits in eastern Washington. Washington State University, Extension Bulletin 419.

Cytokinins + gibberellins — Accel brand of N-(phenylmethyl)-1H-purine-6-amine + gibberellins A_4A_7

Accel is applied in the post-bloom period and is used to chemically thin and improve fruit size in apples. One or two applications of Accel are made when king fruitlets are approximately 5 to 10 mm in diameter (7 to 21 days after full bloom), depending upon cultivar and local conditions. Concentrations vary from 18 to 53.5 fluid ounces per acre. Spray volume per acre ranges from 50 to 200 gallons per acre. Accel should not be mixed with NAA in a spray thinning program or when temperatures fall below 60° (16°C) or rise above 80°F (26°C). Temperatures from 70° to 75°F (21° to 24°C) are most effective.

Cytokinins + gibberellins — Promalin brand of N-(phenylmethyl)-1H-purine-6-amine + gibberellins A_4A_7

Effects on lateral budbreak, branching. On most cultivars, Promalin can promote lateral budbreak and thus branching on most young apple trees. Branch angles can be improved, as can early cropping. Promalin is often applied as a spot application with a brush using a latex paint mixture. Non-ionic wetting agents can be used at the rate of 0.2 to 0.3 percent and should be added to the tank prior to the addition of Promalin. Typically, the material is applied in the spring and when terminal buds begin to swell but before shoots emerge. The trees' response may depend on growing conditions, rootstock, and cultivar. To prevent damage, Promalin should not be applied to low-vigor trees or when temperatures are below 40°F (5°C) or above 90°F (32°C).

Effects on fruit shape "typiness." A single application of Promalin can be applied from the early king bloom to the early stages of petal fall of the side blossoms to possibly improve apple shape and length, especially on 'Red Delicious.' Rates of 1 to 2 pints per acre are recommended. Promalin is not widely used in California to improve apple shape.

Gibberellins A_4A_7 — Provide brand of gibberellins A_4A_7

Post-bloom sprays of Provide can be used to reduce the development of physiological fruit russeting in apples. It should be used in years favorable for russet development and is thought to be most effective on russet that develops from climatic factors rather than physical abrasion. Climatic factors such as precipitation, humidity, and temperature during early fruit growth and development are thought to aggravate fruit russet.

Reference

Washington State University. 1990. Spray guide for tree fruits in eastern Washington. Washington State University, Extension Bulletin 419.

AVG — ReTain brand of [S]-trans-2-amino-4-(2-aminoethoxy)-3-butenoic acid

AVG inhibits the endogenous production of ethylene in plant tissues. Ethylene affects plant processes such as fruit maturation, ripening, and abscission of organs. ReTain may delay maturity in apples, allowing for delayed harvest. The benefits of improved harvest management through use of ReTain may include reduced preharvest fruit drop, enhanced natural red fruit color, and increased size while maintaining fruit firmness and delaying the harvest date. The harvest delay may allow growers to reorganize the harvest of treated blocks and so may improve efficiency with regard to labor management. ReTain is applied four weeks prior to anticipated harvest, and thorough coverage is recommended. To obtain optimum response, either Silwet L-77 or Sylgard organosilicone surfactants must be mixed and sprayed with ReTain. Surfactant concentrations range from 0.05 to 0.1 percent v/v. Apply ReTain under slow drying conditions and maintain a solution pH of 6 to 8.

Apogee brand of prohexadione-calcium, [calcium 3-oxido-5-oxo-4-propionylcyclohex-3-enecarboxylate]

Apogee (BASF Corp., Research Triangle Park, NC) is labeled for use on apples in California, and work is ongoing with other fruits in the hope of securing additional use labels. Apogee is applied as a spray and is useful for reducing shoot growth and fireblight. Typical applications are made in the spring at 1 to 3 inches of shoot growth. Multiple applications and later applications made during the growing season may also be effective for trees with varying degrees of vigor and varying crop loads. A nonionic surfactant is often recommended with an Apogee application. In addition, where water is high in calcium carbonate, a high-quality grade of ammonium sulfate (1 pound per pound of Apogee) is also recommended as an additive to the spray. As application volumes increase from 100 gallons per acre upward, the amount of Apogee should increase per acre to maintain the appropriate concentration in the tank. One effect of Apogee treatment may be increased flowering the following season.

Reference

Costa, G., C. Andreotti, F. Bucchi, E. Sabatini, C. Bazzi, S. Malaguti, and W. Rademacher. 2001. Prohexadione-Ca (Apogee): Growth regulation and reduced fire blight incidence in pear. HortScience 36:931–933.

Dormex — hydrogen cyanamide

Delayed foliation, a long flowering period, and low or weak budbreak are symptoms associated with apples growing under less than adequate chilling conditions. You can use Dormex on apples to enhance budbreak and make flowering more uniform. By making flowering more uniform or altering the flowering of pollinators, you may be able to force fruiting to become more consistent and uniform. Dormex can also stimulate earlier flowering, which may advance the harvest date, especially in warmer regions.

Dormex applications are not suggested until all pruning has been completed. To promote uniform budbreak, apply Dormex as a 4 percent (v/v) spray in not more than 200 gallons per acre combined with a nonionic surfactant not to exceed 0.5 percent (v/v). Dormex should be applied at 30 days before normal bud swell or 35 days before normal budbreak. In many instances, you can shorten the bloom period by several days. The compressed bloom period may reduce vulnerability for fire blight infection. In addition, if more uniform flowering leads to more uniform fruit development, that may improve the success of chemical thinning sprays besides providing more uniform fruit maturity at harvest. Dormex cannot make up entirely for a lack of sufficient chilling hours. A minimum amount of chilling (375 to 500 chill hours at or below 45°F [7°C]) must still accumulate before Dormex application.

Pears (Bartlett, Bosc, and D'Anjou)

Fruitone N, Liqui-Stik brands of NAA — 1-naphthaleneacetic acid, sodium salt; 1-naphthaleneacetic acid, potassium salt, respectively

NAA prevents the preharvest drop of pear fruit. It has been an effective and widely used plant growth regulator for more than 50 years. Sprays are applied either when the first sound uninjured pear fruit drops or 5 to 10 days before harvest in order to prevent the natural "loosening" of the stem from the spur. NAA is often applied at prescribed fruit firmness. The effect of reducing fruit drop may last for as many as 30 days, but the current label specifies 14 days of effective drop reduction. FruitoneN brand should not be sprayed more than twice and applications should not be made closer than 5 days prior to harvest. Liqui-Stik brand becomes effective within 3 to 4 days after application and should not be applied more than two times. These compounds appear to be less effective in cool (60°F [16°C]) than in warm weather (80°F [27°C]). Excessive applications may cause blossom end or core breakdown and premature yellowing and softening of pears. NAA will hold pears on the tree until they become overripe, so picking must be completed soon enough to maintain acceptable firmness for shipping or canning. Excessive applications may reduce flowering in the following season.

References

Allen, F. W., and A. E. Davey. 1945. Hormone sprays and their effect upon the keeping quality of Bartlett pears. Berkeley: University of California Agricultural Experiment Station. Bulletin 692.

Batjer, L. P., A. H. Thompson, and F. Gerhardt. 1948. A comparison of NAA and 2,4-D sprays for control of preharvest drop of Bartlett pears. J. Amer. Soc. Hort. Sci. 40:49–53.

Washington State University. 1990. Spray guide for tree fruits in eastern Washington. Washington State University, Extension Bulletin 419.

Amid-Thin (Amide) brand of NAAmide or NAD — 1-naphthaleneacetamide

Amide has sometimes been used as a chemical thinner for pears. Sprays may be applied at petal fall or within 5 to 7 days after petal fall. Concentrations of 10 to 50 ppm are recommended. It reduces total crop but increases fruit size. Sprays are made about 5 days after petal fall. Earliest sprays cause greater thinning and less foliar injury than later applications. In California, chemical pear thinning is not a common practice.

Reference

Tukey, R. B., and M. W. Williams. 1972. Chemical thinning of Bartlett pears. Washington State University, Extension Manual 3516.

Apricots, Nectarines, Peaches, Plums, and Prunes

ProGibb 4% brand of gibberellins (another product, Ralex, is no longer available)

Gibberellins are used in apricot, nectarine, peach, plum, and 'French' prune to improve fruit firmness, delay harvest, and chemically thin fruit. The last use is relatively new and limited and not currently labeled.

Preharvest use. For use prior to harvest to improve firmness, applications should be made to trees more than 5 years old and in concentrations ranging from 16 to 32 grams per acre. The spray volume should not be less than 100 gallons per acre and sprays should not be made within 5 days of harvest. Applications are made 1 to 4 weeks prior to anticipated harvest (label indicates 3 to 4 weeks). Preharvest applications have been very effective with 'Patterson,' 'Royal/Blenheim,' 'Improved Flaming Gold,' 'Modesto,' and 'Katy' cultivars. Results with peaches, nectarines, and plums are generally positive but vary with cultivar and season. Work with 'French' prune suggests that applications should be made when a blended fruit sample has a soluble solids value of 12 to 14 percent. Fruit firmness is improved and harvest may be delayed by as much as 7 days.

Thinning use (currently not labeled for this use). Gibberellins should be applied as a single spray from mid-May to mid-July depending upon cultivar. Specific spray timing may vary with seasonal temperatures. The chemical inhibits flower bud formation, thus reducing the number of flowers the following season, and thereby has a fruit thinning effect. The normal doses recommended for spray application range from 16 to 32 grams per acre with spray volumes of not less than 100 gallons per acre. Doses from 18 to 38 grams per acre produce consistent thinning responses without overthinning, depending upon species and cultivar. Effective programs have been developed for cling canning peaches and for apricots. For canning peaches, doses should range from 20 to 32 grams per acre and sprays should begin around June 16 (at lower doses) and finish around July 20 (at higher doses in most years). Sprays for apricot should commence in mid- to late May and be completed by early June. Dosages increase with later spray timings. Apricots appear to be more sensitive than canning peaches, so concentrations for apricots should range from 16 to 26 gram per acre. Cases of over-thinning have been noted, but these have mostly resulted from doses that exceeded 38 grams per acre. Non-performance may be a more common occurrence where doses of 16 to 20 grams per acre are used. Application timing should be based on an integrated use of calendar date, number of nodes produced on developing shoots, and temperature prior to application as it relates to fruit maturity.

References

Southwick, S. M., and R. Fritts, Jr. 1995. Commercial chemical thinning of stone fruit in California by gibberellins to reduce flowering. Acta Hort. 394:135–147.

Southwick, S. M., et. al. 1995. Control of cropping in 'Loadel' cling peach by gibberellin: Effects on flower density, fruit distribution, fruit firmness, fruit thinning, and yield. J. Amer. Soc. Hort. Sci. 120:1087–1095.

Cherries (Sweet)

Dormex — hydrogen cyanamide

Growers use Dormex on cherries to increase budbreak and to promote earlier and more uniform vegetative and reproductive budbreak. Earlier budbreak can lead to advanced fruit ripening and more uniform fruit maturity at harvest. Apply Dormex after all pruning is complete. To promote uniform budbreak, apply Dormex as a 4 percent (v/v) spray in not more than 200 gallons per acre combined with a nonionic surfactant not to exceed 0.5 percent (v/v). Even if trees are so large that you cannot achieve uniform spray coverage with 200 gallons per acre, do not use larger per-acre spray volumes. In such situations, Dormex should not be applied.

Applications on cherries should come 30 days or more before normal budbreak. Dormex cannot make up entirely for lack of sufficient chilling hours. As with apples, a minimum amount of chilling (375 to 500 chill at or below 45°F [7°C]) must still accumulate before Dormex application. Observations and research results suggest that the accumulation of suitable chill hours is very important to ensuring reliable results from Dormex applications. Each cherry cultivar varies slightly in the amount of chilling it requires to break rest, and the amount and pattern of chilling varies with each growing season. You really have to monitor the actual amount of chilling accumulated during the season and then apply Dormex based upon the accumulated amount of chilling.

Ethrel brand of ethephon — (2-chloroethyl) phosphonic acid

Ethephon is used on sweet cherries to loosen fruit as an aid in mechanical harvesting. It is not labeled for use in California.

Reference

Micke, W. C., W. R. Schreader, I. T. Yeager, and E. J. Roncoroni. 1975. Chemical loosening of sweet cherries. Calif. Agric. 29(8):3-4.

Gibberellins (GA), gibberellic acid, gibberellin A_3 (GA_3)

There are several manufacturers and marketers of gibberellins, so no one brand will be represented here. Gibberellins are used in sweet cherries prior to harvest as a way to delay maturity and increase fruit firmness, size, and quality. Normally, a rate of 16 to 48 grams per acre is recommended, and applications are made at "straw color." Occasionally, two or more lower-dose applications can be made. With heavy crop loads the effect is not always noticed.

Cytokinins + gibberellins — Promalin brand of N-(phenylmethyl)-1H-purine-6-amine + gibberellins A_4A_7

Promalin is used to increase branching on young sweet cherries. The compound should be applied to vegetative buds on 1-year old wood when approximately 3 to 8 mm of green leaf tissue is pushing out through bud scales. A typical application is 3 parts paint to 1 part Promalin, and the application is made to those buds in which lateral budbreak is desired.

Reference

Proebsting, E. L., Jr. 1972. Chemical sprays to extend sweet cherry harvest. Washington State University, Extension Manual 3520.

Figs

Ethephon — (2-chloroethyl) phosphonic acid

Ethephon was once registered an labeled for use on figs, but currently it is not. Ethephon advances the early, uniform ripening and harvest of dried figs. It was used on 'Calimyrna' and second-crop 'Mission' figs. Earlier treatment increased premature fruit drop due to "puffball" formation, and this reduced both quality and crop. Figs usually would drop 14 to 21 days after treatment, with prompt pick-up harvest for high-quality dried figs. Thorough spray coverage of leaves and fruit would be necessary for this application.

Olives

Liqui-Stik Concentrate brand of NAA — 1-naphthaleneacetic acid, ammonium salt

NAA is used to chemically thin olive fruit, to promote annual bearing, and to increase olive fruit size. Sprays are best applied when fruit is 1/8 to 3/16 inch (3 to 5 mm) in width, generally 1 1/2 to 2 1/2 weeks (12 to 18 days) after bloom. Sprays applied at time of bloom or 1 week later may eliminate the entire crop of olives, and sprays applied 5 weeks after bloom have little beneficial effect. A thorough wetting of all foliage is necessary for even thinning results. Concentrations of NAA for application are adjusted upward by 10 ppm for each day following full bloom during the thinning period. Weak trees and trees with light crops should not be sprayed. Foliar oil sprays will greatly increase the thinning effects of NAA. Hot and dry conditions during and after application may lead to over thinning. All olive cultivars except 'Sevillano' have been effectively thinned with NAA. In hot weather, NAA may kill some tender shoots, but the damage has no lasting effect and may in fact accentuate the thinning.

References

Hartmann, H. T. 1952. Spray thinning of olive with naphthaleneacetic acid. Proc. Amer. Soc. Hort. Sci. 59:187–195.

Hartmann, H. T., and K. W. Opitz. 1977. Olive production in California. Berkeley: University of California Division of Agriculture and Natural Resources, Leaflet 2474.

Walnuts

Ethephon — (2-chloroethyl) phosphonic acid (registered use)

Ethephon advances the normal harvest date for walnuts by hastening hull split and promoting fruit abscission. Earlier harvest, especially on earlier-maturing cultivars, gives better kernel quality. Ethephon also increases the percentage of walnuts that can be removed by a single shaking (once-over harvest) when applied 10 days prior to the expected normal harvest date.

To advance the harvest date, spray ethephon on walnut trees when the nuts are mature, after the packing tissue in the nuts turns completely brown in 95 to 100 percent of the nuts (the packing tissue is the pithy material that fills the space between and around the kernel halves). If you apply ethephon when the packing tissue is white, the kernels may lose weight or shrivel, reducing the nut quality

Sprays are most effective when applied at temperatures between 60° and 90°F (16° and 32°C). Nuts are usually harvested between 7 and 16 days after treatment, depending on cultivar and weather conditions. Early harvest requires a longer drying time for the nuts but can allow drying and hulling equipment to be used over a longer season for treated and untreated areas of the walnut orchard. Ethephon is effective in both concentrate (100 gallons/acre) and dilute (300 to 500 gallons/acre) sprays. Use 3 to 5 pints of ethephon per acre. Best results are obtained with spray concentrations of 300 to 900 ppm (see chart at http://www.cdms.net). Do not exceed 5 pints of this product per acre per year. Thorough coverage of all nuts is essential because ethephon is not translocated through the leaves to the nuts; it must actually be applied to the nuts. No more than 4 hours should elapse between mixing the spray and application. Suggested rates may cause slight yellowing of leaves and some leaf drop on healthy trees. Low-vigor, stressed, or diseased trees should not be sprayed, as they may exhibit excessive leaf drop, twig dieback, and reduced catkin formation. Do not treat more acreage at one time than can be harvested in a reasonable period. When you use ethephon and then delay harvest beyond its optimal time, you can lose nut quality. Do not harvest walnuts any sooner than 5 days after last ethephon application.

References

Kelley, K., and G. S. Sibbett. 1984. Ethephon as a harvest aid for walnuts. Oakland: University of California Division of Agriculture and Natural Resources, Leaflet 21349.

Martin, G. C. 1971. 2-Chloroethyl phosphonic acid as an aid to mechanical harvesting of English walnuts. J. Amer. Soc. Hort. Sci. 96:434–436.

Martin, G. C., H. Abdel-Gawad, and R. J. Weaver. 1972. The movement and fate of 2-chloroethyl phosphonic acid as an aid to mechanical harvesting of English walnuts. J. Amer. Soc. Hort. Sci. 97:51–54.

Olson, W. H., et al. 1977. Lower ethephon rates effective in Walnut harvest. Calif. Agric. 31(7):6–7

3

Plant Growth Regulators for Grape Production

NICK K. DOKOOZLIAN, Viticulture Specialist,
Department of Viticulture and Enology, University of California, Davis

Three plant growth regulators (PGRs) are currently registered for use in grape production in California: gibberellic acid (GA_3), ethephon, and hydrogen cyanamide. In 2001 a fourth compound, florchlorfenuron, was granted an Experimental Use Permit, with full registration expected in 2004. All four compounds are used in table grape production and GA_3 and ethephon are commonly used in raisin production. GA_3 is also available for limited use on wine grapes in some regions of California under a Special Local Needs Permit.

PGRs played a major role in the development of the modern California table grape industry. Prior to the introduction of GA_3 in the 1960s, seeded cultivars dominated production, primarily because of their naturally large berry size. GA_3 allowed the production of seedless grapes with berries of a size and quality similar to those of seeded cultivars. Seedless cultivars now dominate table grape production in most regions of the world. Ethephon and H_2CN_2 have also made major contributions toward improving table grape production efficiency and fruit quality.

PGRs are also commonly used in raisin production, where GA_3 and ethephon are applied to increase fruit quality. At present the use of PGRs in wine grape production is limited, although there is significant interest among growers in expanding the use of these chemicals to regulate fruit set and crop load in the future.

GIBBERELLIC ACID (GA_3)

The primary use of GA_3 for grape growers is to reduce the fruit set and increase the berry size of seedless table grape cultivars. Treatment rates and timings for these applications are quite specific, depending upon the region, cultivar, and desired effects on berry growth and fruit quality. Due to the variability among table grapes in their response and tolerance to GA_3, specific guidelines must be developed for each cultivar.

Seedless Table Grapes

Cluster elongation. GA_3 may be applied to seedless table grapes several weeks prior to bloom to elongate the cluster stem or rachis. However, studies reported that prebloom GA_3 applications initially accelerate cluster growth, but have no effect on either cluster length or berry set at harvest. Recent work has also shown that prebloom applications are detrimental to vine fruitfulness the year following their application. Although GA_3 is registered for this purpose, the application is generally not recommended.

Berry thinning. GA_3 is applied to nearly all seedless table grape cultivars during bloom or *anthesis* (cap fall) to reduce fruit set or cluster compactness. These applications reduce the berry set of most seedless cultivars 10 to 30 percent, and allow fruit sizing manipulations (such as the use of GA_3 and girdling to increase berry size) to be performed without causing excessive cluster compactness. The specific mechanisms responsible for GA_3-induced berry thinning are not known. In the past it was believed that GA_3 acted as a pollenicide. Recent studies, however, have shown that pollen viability and pollen tube growth *in vivo* are not affected by GA_3 applied at bloom. Weather conditions during and immediately following bloom have a major impact on the efficacy of GA_3 applications. Hot (over 90°F [32°C]), dry conditions favor maximum berry thinning, while cool (below 75°F [24°C]), wet conditions reduce treatment efficacy.

GA_3 application rates and timings that are commonly used for berry thinning in seedless table grape cultivars are presented in Table 7. While up to three bloom applications are allowed under the current registration, it is most common to use two on 'Thompson Seedless' and 'Flame Seedless.' Most studies indicate that single and multiple applications result in similar levels of fruit thinning. In practice, however, multiple applications commonly produce larger berries at harvest than single applications because of their effects on berry cell division and elongation. The first spray is typically applied near 50 percent anthesis or bloom and the second near 85 percent bloom. To determine the bloom stage, do a visual estimation of the percentage of total open (exposed) flowers per vine. In the case of 'Crimson Seedless' and 'Ruby Seedless,' a single application near full bloom is generally recommended as a way to avoid excessive thinning and abnormal fruit development.

Table 7. Common GA_3 treatment regimes for commercially important table grape varieties of California.

	Berry thinning			Berry sizing			
Cultivar	Rate (g GA_3/ac)	Number of applications	Application timing (% bloom)	Rate (g GA_3/ac)	Number of applications	Timing of first application (berry diameter)	Comments
SEEDLESS							
'Autumn Royal'	1 to 2	1	85% bloom	—	—	—	Berry sizing sprays have little effect on berry growth and may reduce return fruitfulness.
'Black Emerald'	1 to 2	1	85% bloom	10	1	Berry set + 10 to 14 days	Berry sizing sprays increase potential for postharvest shatter and reduce return fruitfulness.
'Crimson Seedless'	½ to 1	1	85% bloom	8 to 10	1	Berry set + 10 to 14 days	Berry sizing sprays may have little effect on berry growth and often reduce budbreak and return fruitfulness.
'Flame Seedless'	4 to 12	1 or 2	50% and 85% bloom	32 to 40	2 to 3	7 to 8 mm	Higher rates applied for berry sizing often delay or reduce color.
'Perlette'	—	—		40 to 60	2 to 3	4 to 5 mm	Manual brushing required to reduce fruit set.
'Princess'	1	1	85% bloom	—	1	—	Whole-vine sizing sprays not generally recommended due to reduction of return fruitfulness.
'Ruby Seedless'	1	1	85% bloom	12 to 16	1	Berry set + 10 to 14 days	Higher rates at bloom may result in shot berries and reduced fruitfulness the following year.
'Thompson Seedless'	12 to 16	2 or 3	50% and 85% bloom	40 to 60	2 to 3	4 to 5 mm	Higher rates for berry sizing increase postharvest berry shatter.
SEEDED							
'Emperor'	—	—	—	16 to 20	1	12 to 15 mm (berry set + 14 days)	Treatment recommended to reduce berry shrivel. Applied to clusters and foliage.
'Redglobe'	—	—	—	40 to 50 mg/L applied directly to cluster	1	12 to 16 mm (berry set + 14 days)	Berry sizing treatments (40–50 ppm GA_3) applied directly to clusters. Whole-vine applications (up to 16 g/ac) have limited efficacy and may reduce budbreak and return fruitfulness.

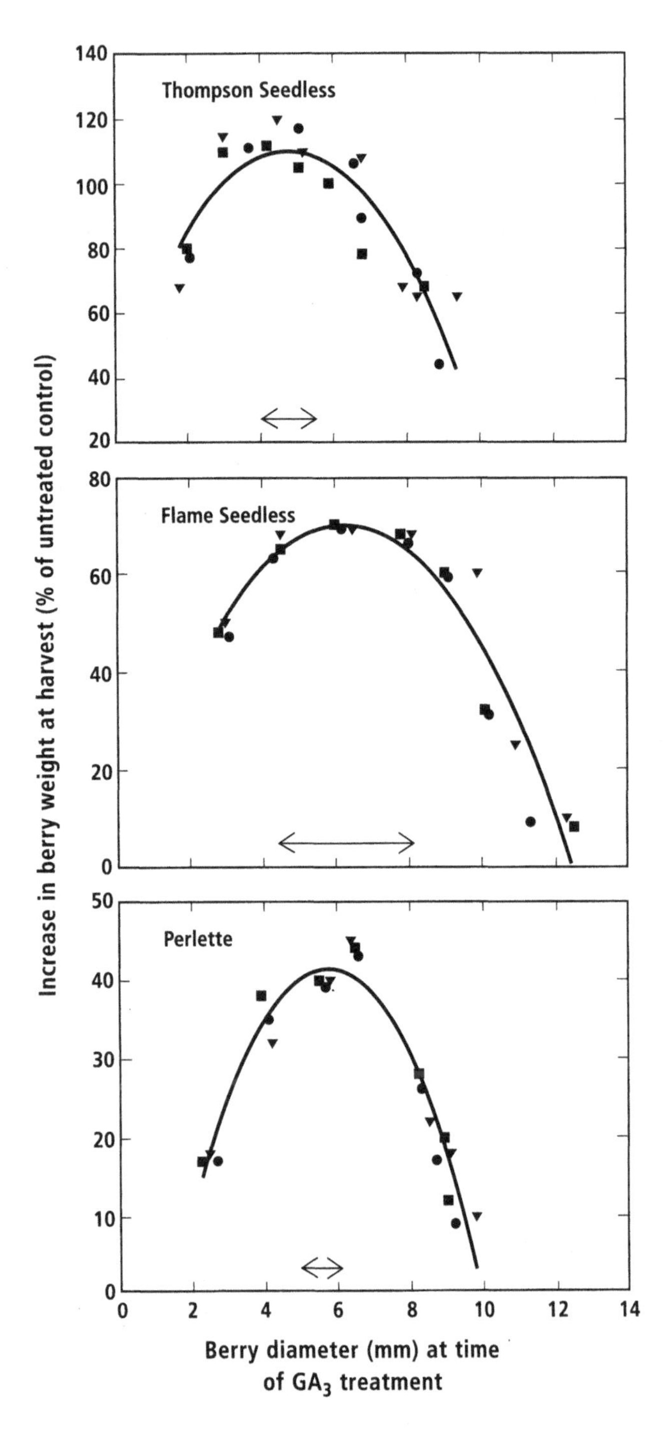

Figure 1. Influence of mean berry diameter at the time of initial GA_3 treatment on the berry weight of 'Thompson Seedless' *(upper graph)*, 'Flame Seedless' *(middle graph)*, and 'Perlette' *(lower graph)* table grapes at harvest. Berry weight is expressed as the percent increase compared to the untreated control. Data points represent the mean of twelve single-vine replicates per application timing in 1992 (▼), 1993 (■), and 1994 (●). Horizontal lines with arrows indicate the window for maxium berry response to GA_3 in each cultivar.

Studies have shown that the effects of GA_3 on fruit thinning are generally similar if applications are performed anywhere between 25 and 100 percent bloom. Applications made during late bloom, however, increase the length and weight of ellipsoidal or cylindrical berries such as 'Thompson Seedless' more than early bloom applications. For this reason, if a single thinning application is used on a seedless cultivar, it is best performed near full bloom.

Berry sizing. GA_3 is applied to seedless table grapes near fruit set to increase berry size. Berry growth is stimulated as a result of increased cell division (increasing the total number of cells per berry) and cell elongation (producing larger cells in the berry). These treatments increase berry size at harvest by 50 percent or more, depending upon the cultivar, application rate, and the number of applications performed. Standard treatment regimes for the major cultivars grown in California are presented in Table 7. When multiple applications are used, the second and third applications are typically performed 4 to 7 days and 8 to 14 days after the first, respectively, depending on the cultivar and region.

Application timing has a major impact on the efficacy of GA_3 applied at fruit set. For most cultivars, including 'Thompson Seedless' and 'Perlette,' the treatment window for the maximum sizing response is quite narrow (Figure 1). In contrast, 'Flame Seedless' has a relatively broad treatment window for maximum response (Figure 1).

While the berry size of most seedless cultivars continues to improve as the amount of GA_3 applied at fruit set is increased, the response generally saturates between 80 and 120 grams per acre. Several problems commonly arise as the amount of GA_3 used for berry sizing increases. First, GA_3 decreases berry attachment to the capstem or pedicel, resulting in significant *berry shatter* (when berries separate from the cluster) during the harvest and postharvest periods. Second, GA_3 delays fruit maturity and reduces berry color (in 'Flame Seedless,' for example), causing harvest to be extended or the total amount of harvestable fruit to be reduced. Third, excessive rates of GA_3 applied for berry sizing may result in reductions in both cluster number per vine and cluster size the year following their application. This reduction in vine productivity can range from 10 to 15 percent for tolerant cultivars such as 'Thompson Seedless,' and up to 50 percent or more for highly sensitive cultivars such as 'Princess.' Growers must therefore compromise between their desire to achieve maximum berry size and the potential for harm to fruit quality and vine productivity when selecting application rates. In addition to applying lower amounts of GA_3, one common strategy when treating highly sensitive cultivars is to delay sizing applications until approximately 2 weeks after fruit set. This allows many of the buds on the vine to complete fruit-bud differentiation for the following season before application, partially reducing potential detrimental effects on return fruitfulness.

Seeded Table Grapes

Berry sizing. The use of GA_3 on seeded table grapes is limited primarily because of its phytotoxic effects on vine growth and productivity. It is generally accepted that seeded grapes are more sensitive than seedless grapes to GA_3 applications because seeded grapes contain higher natural or endogenous levels of gibberellic acid. This may also explain why GA_3 is much less effective at increasing the berry size of seeded grapes than of seedless grapes, particularly when one considers that the seeds provide a source of GA for the berry. Most seeded grapes are quite sensitive to foliar-applied GA_3, and irregular leaf and shoot growth characteristics are commonly observed even at low application rates. Foliar applications performed in the spring result in severe reductions in bud fruitfulness the following year and may even cause bud death. If applied before or during bloom, GA_3 causes abnormal cluster stem and berry development and leads to the formation of *shot berries,* small, underdeveloped berries without seeds that must be removed by hand prior to packing.

GA_3 may be used to increase the berry size of 'Redglobe,' the primary seeded table grape grown in California. This application is particularly effective on berries that contain only one or two seeds, although most 'Redglobe' berries contain three or four. If you want to avoid problems with erratic budbreak and reduced fruitfulness the year after application, foliar sizing sprays cannot exceed 16 grams per acre and must be applied at least 2 weeks after fruit set. Recent work has shown, however, that foliar applications at this rate have little effect on berry growth. In an effort to increase the efficacy of berry sizing treatments, many growers now apply higher rates of GA_3 directly to the clusters either by manually immersing each cluster or by spraying individual clusters using hand-held spray wands and foliage shields to keep the solution away from developing buds. The concentration of GA_3 normally used for this treatment is 40 to 50 mg /L (ppm). GA_3 applied by this method increases the average berry weight of 'Redglobe' by 5 to 10 percent.

GA_3 is also applied to the seeded cultivar 'Emperor' approximately 2 weeks after fruit set to reduce berry shrivel, a physiological disorder that causes the cluster stem to dry and the berries to soften and raisin prior to harvest. This application also increases berry size.

Raisin Grapes

At bloom, GA_3 may be applied to 'Thompson Seedless,' the dominant raisin grape cultivar in California, to reduce fruit set and cluster compactness. The application also increases berry fresh weight and length and results in improved raisin quality as graded by the airstream sorter. This treatment is most applicable in vineyards with a history of poor raisin quality or bunch decay and in seasons with high crop loads or late fruit maturity. A typical treatment regime consists of a single application of 3 to 8 grams GA_3 per acre at 50 to 85 percent bloom. Low rates (4 to 10 grams GA_3 per acre) may also be applied at fruit set to increase the berry size of 'Thompson Seedless' raisin grapes. This application is generally not recommended, though, due to reports that it delays fruit maturation and leads to problems with capstem removal when processing the dried fruit. Growers should check with their raisin packer before applying GA_3 at fruit set.

GA_3 is also commonly applied to the raisin cultivar 'Black Corinth' ('Zante Currant'). The cultivar sets fruit parthenocarpically (without fertilization of the ovary), and produces small, seedless berries that are dried for midget raisins. Its natural berry size is too small for commercial purposes, so GA_3 is used to stimulate berry cell division and increase berry size. GA_3 is typically applied 3 to 5 days after full bloom at 4 to 6 grams per acre.

Wine Grapes

In the early 1970s GA_3 was registered for reducing fruit set and cluster compactness of wine grape cultivars in California. This application was shown to reduce the potential for bunch rot. The registration was removed after a few seasons, however, when growers reported significant decreases in vine fruitfulness the year following GA_3 application. In the mid-1990s this registration was reinstated in several regions of the state under a Special Local Needs Permit. In the northern San Joaquin Valley, for example, 'Zinfandel' vines may be treated with 2 to 4 grams of GA_3 per acre several weeks before bloom to reduce cluster compactness. At present only a small percentage of the state's wine grape acreage is eligible for this treatment.

Most wine grape cultivars, like seeded table grape cultivars, are highly sensitive to GA_3. Rates effective for reducing cluster compactness generally reduce vine fruitfulness the following year. Acceptable rates for this application vary considerably among wine grape cultivars, so specific recommendations must be developed for each cultivar.

ETHEPHON

Table Grapes

Ethephon enhances pigment (anthocyanin) or color accumulation in the skin of grape berries. While this effect is most visible in red cultivars, the color of black cultivars is also improved. In many cases ethephon is applied to advance the date of harvest, particularly for early ripening cultivars such as 'Flame Seedless.' It also improves production efficiency by increasing the amount of fruit that can be harvested in the first picking, thus reducing the total number of pickings necessary. The application of ethephon partially overcomes the effects of berry sizing manipulations, such as GA_3 and trunk girdles applied at fruit set, which are generally detrimental to fruit color. The specific mechanisms involved in ethephon-induced color and ripening enhancement in grape, a non-climacteric fruit, are unknown, but researchers believe the mechanisms of action are related to the compound's ethylene-evolving properties.

For best results on early or mid-season ripening cultivars such as 'Flame Seedless,' ethephon should be applied shortly after berry softening (initiation of fruit ripening), the moment when 5 to 10 percent of the berries appear colored. In early fruit ripening districts such as the Coachella Valley, growers often apply trunk girdles in conjunction with ethephon treatment at berry softening to further hasten fruit coloration and maturation. Ethephon may be applied to late-ripening cultivars such as 'Crimson Seedless' or 'Christmas Rose' immediately after berry softening as defined above (5 to 20 percent of the berries showing color). If the grower wants to delay harvest or fruit coloration, ethephon can be applied to these cultivars as late as 4 to 6 weeks before the anticipated harvest date. The application rate on table grapes is 1 to 2 pints of formulated product per acre (21.7 percent active ingredient [*a.i.*]).

Ethephon reduces berry firmness, particularly when it is applied near the time of fruit softening. This reduction is rate-dependent and most noticeable on cultivars in which natural berry firmness is low, such as 'Ruby Seedless' or 'Redglobe.' Fruit soluble solids are typically not affected or are increased only slightly, while acidity usually is reduced.

Raisin Grapes

Ethephon advances the maturity of 'Thompson Seedless' and 'Fiesta' approximately 7 days when applied near fruit softening (approximately 5 percent of berries soft). This allows growers to harvest at the desired maturity level 1 week earlier than normal, thus reducing the risk that fall rains will ruin the raisin crop, or to harvest on the regular date at a higher maturity level in order to improve raisin quality. The application rate is 1 to 2 pints of formulated product per acre (21.7 percent a.i.).

HYDROGEN CYANAMIDE (H_2CN_2)

The commercial use of hydrogen cyanamide (H_2CN_2) in California is generally limited to early season table grapes grown in the Coachella Valley of Riverside County. Grapevines grown in this region receive inadequate winter chilling for optimum budbreak in the spring. This causes budbreak to be erratic and delayed and reduces the number of shoots and clusters per vine. Treatment with H_2CN_2 after pruning can overcome the effects of inadequate winter chilling on vine growth. Total budbreak is increased and growth uniformity improved. Depending upon the date of pruning and H_2CN_2 application, budbreak and harvest can be significantly advanced. For best results, winter pruning should be delayed as long as possible in order to maximize chilling exposure. In many cases, however, vines must be pruned and treated with H_2CN_2 prior to desired chilling exposure in order to initiate budbreak and advance the harvest date into the desired market window. Depending on temperatures following application, budbreak typically occurs 20 to 30 days after H_2CN_2 treatment. Fruit harvest is advanced 1 to 3 weeks over untreated vines, depending on temperatures following budbreak.

In the Coachella Valley, vines are usually treated with a 2 percent (v/v) solution of active ingredient per acre (a maximum 2 gallons of a.i. may be applied per acre) within 72 hours after pruning. Safety precautions must be carefully followed. H_2CN_2 is a restricted-use material and requires a closed system for mixing. Care should be taken to prevent spray drift from reaching surrounding crops. H_2CN_2 drift causes significant defoliation and potential crop losses in lemons.

This compound is registered for use on table, raisin, and wine grapes throughout the state. However, because most regions of California receive ample winter chilling (400 hours or more below 45°F [7°C]) for normal growth, there is little economic reason to use it outside of the Coachella Valley. The erratic or incomplete budbreak of grapevines grown in other regions of the state is not believed to be the result of inadequate winter chilling. While vines treated with H_2CN_2 initiate growth sooner than untreated vines, studies show that this does not consistently advance fruit maturity or improve developmental uniformity. This is likely due to the lower rate of degree-day accumulation in these regions, particularly in the early spring, compared to the Coachella Valley.

FORCHLORFENURON

Forchlorfenuron (N-(2-chloro-4-pyridyl)-N'-phenylurea; commonly known as CPPU) is a synthetic cytokinin that has significant physiological activity on many fruits, including grapes. The compound was granted an Experimental Use Permit for grapes in California in 2001, and full registration is expected by 2004.

The primary physiological effects of forchlorfenuron on grapevines involve the regulation of fruit set, berry growth, and berry development in table grapes. When applied prior to or during anthesis or bloom, forchlorfenuron increases the fruit set of both seedless and seeded grape cultivars. When applied following fruit set, forchlorfenuron stimulates cell division and cell elongation, resulting in increased berry size. These applications can also delay fruit maturation, slowing the accumulation of sugar and color and the respiration of organic acids. High rates applied at fruit set may permanently retard color development, particularly on sensitive cultivars such as 'Flame Seedless,' and have also been reported to alter berry flavor and texture. Another potential problem involves the alteration of berry shape, particularly on ellipsoidal or cylindrical berries such as 'Thompson Seedless' or 'Crimson Seedless.' Forchlorfenuron stimulates periclinal cell division in the berry, leading to rounder or more oval berries than when GA_3 is used alone.

Forchlorfenuron is commonly applied at fruit set in conjunction with GA_3 to increase the size and firmness of the standard seedless table grape cultivars including 'Thompson Seedless' and 'Flame Seedless.' Optimum rates vary depending upon the cultivar, district, and desired maturation delay, ranging between 3 and 12 grams per acre. Rates lower than 3 grams per acre do not appear to be effective, while rates exceeding 8 grams per acre delay maturity and significantly inhibit color development in most red and black cultivars. In general, a range of between 4 and 6 grams per acre is recommended for red and black cultivars while slightly higher rates may be used for white cultivars when a delay in fruit maturation rate is desired.

Forchlorfenuron may also be used as a sizing spray for GA_3-sensitive seedless (e.g., 'Ruby Seedless') and seeded (e.g., 'Redglobe') cultivars in which whole-vine sprays of GA_3 are detrimental either to fruit quality or to vine fruitfulness. Forchlorfenuron does not reduce the fruitfulness of either seedless or seeded table grape cultivars the year following its application, and no negative effects on vegetative growth or development have been reported.

References

Christodoulou, A. J., R. J. Weaver, and R. M. Pool. 1968. Relation of gibberellin treatment to fruit set, berry development, and cluster compactness in Vitis vinifera grapes. Proc. Am. Soc. Hort. Sci. 92:301–310.

Dokoozlian, N. K., D. Luvisi, M. Moriyama, and P. Schrader. 1995. Cultural practices improve the color and size of 'Crimson Seedless' table grapes. Cal. Agric. 49:36–40.

Dokoozlian, N. K., M. M. Moriyama, and N. C. Ebisuda. 1994. Forchlorfenuron increases the berry size and delays the maturity of 'Thompson Seedless' table grapes. Proc. Inter. Symp. on Table Grape Prod. June 28 and 29. 1994. Anaheim, CA. pp. 63–68.

Dokoozlian, N. K., L. E. Williams, and R. A. Neja. 1995. Chilling exposure and hydrogen cyanamide interact in breaking dormancy of grape buds. HortSci. 30:1244–1247.

Jensen, F. L. 1975. Effect of ethephon on color and fruit characteristics of 'Tokay' and 'Emperor' table grapes. Am. J. Enol. Vitic. 26:79–80.

Lynn, C. D., and F. L. Jensen. 1966. Thinning effects of bloomtime gibberellin sprays on 'Thompson Seedless' table grapes. Am. J. Enol. Vitic. 17:283–289.

Weaver, R. J., A. N. Kasimatis, and S. B. McCune. 1962. Studies with gibberellin on wine grapes to decrease bunch rot. Am. J. Enol. Vitic. 13:78–82.

Weaver, R. J., and S. B. McCune. 1961. Effect of gibberellin on vine behavior and crop production in seeded and seedless Vitis vinifera. Hilgardia:30(15).

4

Plant Growth Regulators for Vegetable Crops

MIKE MURRAY, Vegetable Crops Advisor, Colusa and Glenn Counties, and
LANCE BEEM, Field Market Development Specialist, Valent USA

The use of plant growth regulators (PGRs) to manipulate flowering patterns, reproductive cycles, crop maturity, in-storage sprouting, and plant architecture helps fresh market, processor, and seed producers achieve their objectives in a coordinated and efficient manner. At this time, the use of PGRs to achieve these goals in vegetable crops is limited to just a few examples. Additional future applications and registrations are likely as growers and researchers identify expanded uses for existing materials as well as new products.

Only PGRs that have a current registration in the state of California at the time of this publication will be discussed here. A number of PGRs that are used in other states or countries are omitted. Additional products are marketed under such categories as "biostimulants" and "growth enhancers," but those will not be discussed here. Such products are not registered as PGRs, though their manufacturers may claim that they improve plant growth or yield. They typically lack unbiased data to substantiate their usefulness, are of an undetermined or poorly quantified chemical composition, or have not been submitted to the same degree of scrutiny as registered PGRs. Materials such as mixed cytokinins and organic acids are included in this category.

Also omitted from the present discussion are two vegetable PGR usages: ethylene gas used for postharvest ripening of tomatoes or melons and chloropropham-isopropyl-m-chlorocarbanilate (CIPC) used to prevent or delay sprouting in stored potatoes. These materials are applied under specialized conditions and most pest control advisers (PCAs) will not be involved with them. Those who will be working with these applications should obtain specialized assistance.

Our discussions are organized by crop or commodity, and each listing consists of a short statement describing the biological effects of the product, a brief description of important parameters affecting the results obtained, the identification of any cautions associated with using the product, and selected references for obtaining additional information. Information presented in Chapter 1 related to PGR modes of action and biological effects will not be repeated here, so a thorough basic understanding of PGRs is required before you proceed. The responsibility for confirming that a specific PGR is registered for use on a specific commodity rests with the PCA making the recommendation. Introductions of new registrations and expirations of old ones are constantly occurring. For specific information on registered uses, rates, critical timings, and precautions for specific PGRs, contact the manufacturer's representative or access product information through the Internet. Several manufacturers maintain these Web sites. Crop Data Management Systems (CDMS) lists most PGR labels on its site (http://www.cdms.net).

Our intent in furnishing the following information is to provide you with an inventory of currently registered PGR applications. We do not discuss the extent to which these materials are commonly used by producers. Some uses such as ethephon for hybrid squash seed or processing tomato production are widespread, while other uses such as ethephon on cantaloupes or peppers are rare, owing to potential decreases in crop quality.

Artichokes

Gibberellic acid (GA_3)

When applied properly, GA_3 can increase earliness and uniformity of bud development without causing significant damage to plant development or yield. Applications to annual crops can advance harvest by as much as 8 weeks sooner than the harvest date for untreated plants.

For annual production, two or three GA_3 applications applied at 2-week intervals beginning 5 to 7 weeks after transplanting provide the maximum growth response. Applications to annual crops within 10 weeks of harvest do not promote earliness or uniformity. Perennial crops should receive a single application approximately 6 weeks before the anticipated harvest date. The foliage near the growing point should be wetted to the point of runoff, when possible.

Improper GA_3 application rates or timing can reduce plant vigor, increase susceptibility to black tip or spider mite damage, cause early buds to be misshapen, or lead to brittle leaves. This misapplication may include premature applications, excessive GA_3 rates, or high air temperatures at the time of application.

References

Schrader, W. L. 1993. Growth regulator gives earlier harvests in artichokes. Calif. Agric. 48(3):29–32.

Schrader, W. L., and K. S. Mayberry. 1997. Artichoke production in California. Oakland: University of California, Division of Agriculture and Natural Resources, Publication 7221.

Snyder, M. J., N. C. Welch, and V. E. Rubatzky. 1971. Influence of gibberellin on time of bud development in globe artichoke. Hort Science 6:484.

Cantaloupes

Ethephon ([2-chloroethyl] phosphonic acid)

Ethephon can hasten the maturity of field-grown cantaloupes by promoting fruit abscission (*slipping*). This permits more uniform maturity and harvest. Good foliar spray coverage is required to attain maximum benefits. Producers should be prepared to harvest treated fields within 2 to 5 days of treatment, as field holding periods are decreased.

Fruit quality, in terms of soluble solids and flesh color, does not improve after harvest. For this reason you should not apply ethephon until the fruit has developed marketable levels of soluble solids and color. Ethephon promotes the abscission of immature fruit as well as marketable fruit. If ethephon is applied too early, the crop may have poor flesh color development or substandard soluble solids.

Crop phytotoxicity may occur if air temperatures are high (>95°F[>35°C]) at treatment or shortly thereafter. Proper timing of applications is critical for ethephon, but the optimal timing may change during the season. Do not treat fields if night temperatures fall below 60°F (16°C). Do not treat stressed plants or those with split fruit-sets.

References

Hartz, T. K., K. S. Mayberry, and J. Valencia. 1996. Cantaloupe production in California. Oakland: University of California, Division of Agriculture and Natural Resources, Publication 7218.

Pratt, H. K., J. D. Goeschl, and F. W. Martin. 1977. Fruit growth and development and the role of ethylene in the 'Honey Dew' muskmelon. J. Amer. Soc. Hort. Sci. 102:203–210.

Celery

Gibberellic acid (GA_3)

The use of GA_3 can result in increased plant height or early yield, under winter growing conditions, and can reduce plant stress caused by saline soils. A single application 1 to 4 weeks prior to harvest has produced the most consistent results. California product label restrictions require ground application. Applications more than 4 weeks before harvest may result in excess bolting (seed stalk formation), reducing product quality.

Reference

Koike, S. T., K. F. Schulbach, and W. E. Chaney. 1996. Celery production in California. Oakland: University of California, Division of Agriculture and Natural Resources, Publication 7220.

Cucumbers and Squash

Ethephon

The timely application of ethephon to cucumbers or squash results in the production of additional pistillate (female) flowers and the suppression of staminate (male) flowers. Inexpensive and efficient hybrid squash seed production results when only the female parent rows are treated. Multiple applications at early plant growth stages are required. Treatment also advances the initiation of flowering, so care must be exercised to ensure that both hybrid parents flower at the same time.

Ethephon may also be applied to fresh market pickling cucumbers to cause additional female flower formation and higher fruit yields. One in every ten rows should be left untreated to ensure that there will be enough male flowers for pollination.

The results obtained from ethephon applications are temporary; multiple, sequential applications during early plant growth are necessary for optimum results. The specific variety being grown, growing practices, environmental conditions, and geographic location affect the results of this treatment. There is no "standard practice" that is best for all cultivars, and small, experimental applications should be made to unfamiliar varieties before large acreages are treated.

Crop phytotoxicity, distorted floral or vegetative growth or inhibition of plant growth may result from excessive chemical rates, treatment of stressed plants, or applications that are followed by high air temperatures (>95°F [>35°C]). These symptoms are transitory, but flowering patterns may be altered and poor pollination may result.

References

McMurray, A. L., and C. H. Miller. 1968. Cucumber sex expression modified by 2-chloroethanephosphonic acid. Science 162:1397–1398.

Murray, M. 1987. Field applications of ethephon for hybrid and open-pollinated squash (Cucurbita spp.) seed production. Acta Horticulturae 201:149–156.

Murray, M., T. K. Hartz, and K. Bradford. 1997. Cucurbit seed production in California. Oakland: University of California, Division of Agriculture and Natural Resources, Publication 7229.
Sims, W. L., and B. L. Gledhill. 1969. Ethrel effects on sex expression and growth development in pickling cucumbers. Calif. Agric. 23(2):2–3.

Lettuce (Seed)

Gibberellic acid (GA_3)

Timely applications of GA_3 are used to induce uniform bolting and increased seed production in tight-headed (iceberg) lettuce. Three treatments of 10 ppm GA_3 should be applied, at the fourth-, eighth-, and twelfth-leaf stages. These treatments should be supplied in sufficient water to thoroughly wet the plant foliage.

References

Harrington, J. F. 1960. The use of gibberellic acid to induce bolting and increase seed yield of tight-heading lettuce. Proc. Amer. Soc. Hort. Sci. 75:476.

Onions

Maleic hydrazide (MH)

Maleic hydrazide is used to prevent or delay postharvest sprouting. Apply MH when bulbs are fully mature and have 5 to 8 green leaves. The onion necks must be soft enough to fall, if they have not already done so. A typical application timing would be when 50 percent of the tops have fallen but all of the tops are still green. Green tops are essential for chemical absorption. If onions are sprayed more that 2 weeks before harvest they may develop spongy bulbs. Do not apply MH to onions with poor keeping quality: the chemical will not make poor-keeping varieties equal to good-keeping varieties.

Factors that may result in erratic or poor results include treatment of stressed plants, applications within 24 hours of overhead irrigation or rain, and application shortly before temperatures rise above 85°F (29°C).

Peppers

Ethephon

Late-season applications of ethephon can accelerate fruit coloring and maturity and to improve uniformity. Apply ethephon to bell peppers when 10 percent of the fruit are red or chocolate, and to chili or pimento pepper varieties when 10 to 30 percent are red or chocolate. Ethephon applied too early or under adverse conditions may cause pale red fruit and will not ripen immature green fruit. The holding ability or "ship-ability" of treated fruit may be poorer than for untreated fruit.

Ethephon applications made under high air temperatures (95°F [35°C]) or plant stress conditions due to other factors may result in plant damage, including defoliation, sunburn, and shedding of immature fruit. Plants that are subjected to such applications will not remain vigorous. Fruit may not be harvested within 5 days of an ethephon application.

References

Hartz, T. K., M. LeStrange, K. S. Mayberry, and R. F. Smith. 1996. Bell pepper production in California. Oakland: University of California, Division of Agriculture and Natural Resources, Publication 7217.
Osterli, P. P., R. M. Rice, and K. W. Dunster. 1975. Effect of ethephon on bell pepper ripening. Calif. Agric. 29(7):3.
Sims, W. L., D. Ririe, R. A. Brendler, M. T. Snyder, D. N. Wright, V. H. Schweers, and P. O. Osterli. 1974. Factors affecting ethephon as an aid in fruit ripening of peppers. Calif. Agric. 28(6):3–4.
Smith, R., T. Hartz, J. Aguiar, and R. Molinar. 1998. Chile pepper production in California. Oakland: University of California, Division of Agriculture and Natural Resources, Publication 7244.

Potatoes

Maleic hydrazide (MH)

Maleic hydrazide is used on potatoes to control tuber sprouting in storage. It must be applied in the field after tubers are formed, while vines are still green and healthy. This treatment will not make a poor-holding variety into a good-holding one.

For best results, MH applications should be made 3 to 4 weeks before harvest, but at least 2 weeks prior to vine dessication. Applications made more than 4 weeks before harvest may result in crop damage or small tubers.

Gibberellic acid (GA_3)

GA_3 is used on seed potatoes to stimulate uniform sprouting and to break dormancy in seed potatoes that have not had a full rest period. Do not treat plants that have been stressed during the growing cycle. Do not use on crops that are grown for seed.

The seed potatoes should be dipped in a GA_3 solution. Determine the appropriate concentration of GA_3 on the basis of anticipated soil temperatures. GA_3 may increase the stem's susceptibility to Rhizoctonia by producing elongated underground stems. GA_3 may also cause abnormal stems, stolons, or elongated tubers and promote the production of excessive numbers of tubers.

References

Kim, M. S. L., E. E. Ewing, and J. B. Sieczka. 1972. Effects of chloropropham (CIPC) on sprouting of individual eyes on plant emergence. Amer. Potato J. 49:420–431.

Timm, H., C. Bishop, and B. Hoyle. 1959. Investigations with maleic hydrazide on potatoes. 1. Effect of time of application and concentration upon potato performance. Amer. Potato J. 36:115–123.

Timm, H., L. Rappaport, P. Primer, and O. E. Smith. 1960. Sprouting, plant growth, and tuber production as affected by chemical treatment of white potato seed pieces. II. Effects of temperature and time of treatment with gibberellic acid. Amer. Potato J. 37:357–365.

Rhubarb

Gibberellic acid (GA_3)

GA_3 may reduce or partially replace the chilling requirement for forcing rhubarb. A single application is made to each individual crown. If temperatures in the forcing houses rise above 50°F (10°C) immediately following a GA_3 application, yield losses or poor stalk color may result.

References

Tompkins, D. R. 1966. Rhubarb petiole color and forced production as influenced by gibberellin, sucrose, and temperature. Proc. Amer. Soc. Hort. Sci. 89:472.

Tompkins, D. R., and M. A. Caldwell. 1972. Rhubarb production. USDA Leaflet 555.

Tomatoes (Processing)

Ethephon

Foliar applications of ethephon 3 to 4 weeks before anticipated harvest can advance the maturity and improve the color uniformity of processing tomatoes. Depending on variety, climate, ethephon rate, and other variables, harvest may be advanced from 7 to 10 days over untreated crops.

Thorough foliar spray coverage is essential, and ground applications are typically more effective than aerial applications. Applications made under high temperatures (>95°F [>35°C]) may cause leaf damage, defoliation, or sunburnt fruit. Appropriate ethephon rates change throughout the season. Early season applications under high air temperatures may be 25 percent of the rates used later in the season, when days are short and temperatures lower.

Applications should be timed according to the maturity of the specific field being treated. To determine field maturity, remove several plants and divide the fruit into red, mature green, and immature green. Apply ethephon when 5 to 15 percent of the fruit are red or breaker and a minimum amount of fruit are immature green. Immature green fruit will not ripen after being treated with ethephon.

Treated fields must be harvested in a timely manner since treated fruit does not have the field-holding capacity of untreated fruit. If plant damage results from ethephon applications, an earlier harvest may be required in order to minimize fruit sunburn and color degradation. Processors sometimes express concerns about ethephon's negative effects on fruit color. This problem may be more accurately associated with defoliation and sunburn rather than with ethephon, and further emphasizes the need to maintain a healthy plant.

Tomatoes (Fresh)

Ethephon

Many of the same comments made above concerning the use of ethephon on processing tomatoes apply also to fresh tomatoes. There are, however, important differences between the two crops. Ethephon typically would not be applied to fresh market tomatoes until 3 to 6 days before the anticipated harvest. Once ethephon is used, the option to conduct additional harvests is lost, as the plants will become senescent. Do not apply ethephon to greenhouse tomatoes.

References

Hartz, T. K., and G. Miyao. 1997. Processing tomato production in California. Oakland: University of California, Division of Agriculture and Natural Resources, Publication 7228.

Iwahori, S., and J. M. Lyons. 1969. Accelerating tomato fruit maturity with ethrel. Calif. Agric. 23(6):17–18.

Miller, C. H., R. L. Lower, and A. L. McMurray. 1969. Some effects of ethrel (2-chloroethanephosphonic acid) on vegetable crops. Hort. Science. 4:248–249.

Robinson, R. W., H. Wilczynski, F. G. Dennis, Jr., and H. H. Bryan. 1968. Chemical promotion of tomato fruit ripening. Proc. Amer. Soc. Hort. Sci. 93:823–830.

5

Plant Growth Regulators for Field Crops

WARREN E. BENDIXEN, Farm Advisor, Santa Barbara and San Luis Obispo Counties, and
BILL L. WEIR, Field and Vegetable Crops Advisor, Merced County

Plant growth regulator (PGR) use on field crops has generally been limited to a few crops because of a low cost-benefit ratio and the relatively low value of field crops. PGRs are currently used to manipulate crop maturity, reproductive cycles, and plant height, and to strengthen vegetative structures. In some crops, breeding programs have been very successful in changing the plant harvest index and structural strength, thereby reducing the need for PGRs. One example is new, shorter, stiff-straw wheat varieties. New developments through biotechnology and advanced breeding programs will continue to modify plants and impact the relative value of PGRs, but continued advances in PGR development are also likely to sustain interest in PGRs as a management tool.

This chapter deals with PGRs with current registration on field crops in California. Continued research may lead to the use of additional PGRs for field crops. Growers and pest control advisers (PCAs) need to verify that a PGR is currently registered in California on the desired crop before they recommend its use. Manufacturers may claim that some products improve plant growth or serve as growth enhancers, but if they are not registered as such they will not be discussed in this chapter. Some of these non-PGR materials may be classified as fertilizers based on their chemical analysis. Brief descriptions of important information affecting crop responses and crop safety are included for each PGR-crop combination below.

The three plant growth regulators registered for field crops in California are ethephon, gibberellic acid (GA), and mepiquat chloride. These products will be covered as they relate to use on specific crops. The same PGR may be registered under two separate names: for example, ethephon is registered as Prep on cotton and as Cerone on wheat and barley.

Barley and Wheat

Ethephon ([2-chloroethyl] phosphonic acid)

Ethephon, marketed as Cerone for wheat and barley, is registered as an anti-lodging agent. To assure safe and effective use of Cerone, applications should follow the directions on the label. High air temperatures following Cerone applications may cause crop injury and reduce yields.

Cerone should be applied at crop growth stages between the flag leaf and late boot (Feekes scale 8 to 10). Cerone applications can reduce lodging in wheat varieties by reducing plant height and stiffening stems. In a California trial on the 'Serra' wheat variety, Cerone increased grain yields more than 1,000 pounds per acre. Trials on prevention of lodging have been conducted for some California barley varieties, but the results have not been consistent.

Crop production practices that increase plant height, plant density, and grain yield also increase the lodging hazard. Grain yield, bushel weight, and kernel size decrease when lodging occurs. Harvesting costs increase as the severity of the lodging increases, and during harvest more grain is lost on the ground in lodged fields.

References

Bendixen, W. E. 1993. Ethephon increases the yield of 'Serra' wheat. Proc. Plant Growth Regulator Soc. Amer.

Bendixen, W. E. 1994. Ethephon—a plant growth regulator—increases the yield of 'Serra' wheat. Proc. Plant Growth Regulator Soc. Amer.

Bendixen, W. E., T. E. Kearney, and J. F. Williams. 1990. Cerone applications on 'Serra' wheat. Proc. West. Plant Growth Regulator Soc.

William, J., L. Jackson, and J. Webster. 1992. Effect of Cerone plant growth regulator on 'Serra' wheat. Proc. West. Plant Growth Regulator Soc.

Cotton

The two most common plant growth regulators used in California cotton production are ethephon (Prep) and mepiquat chloride (Pix).

Ethephon ([2-chloroethyl] phosphonic acid)

Ethephon, marketed as Prep in cotton production, assists in boll opening. To assure safe and effective use of Prep, applications should follow the directions on the label.

A properly timed application of ethephon can increase the rate of maturation and opening of mature cotton bolls. The treatment should not be applied too early in boll maturation, however, since that would increase the rate of abortion of immature bolls. Early research on cotton in California's San Joaquin Valley and in other states showed that applying ethephon 1 week before applying defoliants gave the best results. Since defoliation should commence when the crop has four harvestable bolls above the topmost cracked boll, ethephon should be applied about 1 week before this time. The higher the ethephon application rate, the less time you have to wait after application before you can begin defoliation treatments. To avoid making an extra trip across the field, though, it has become common practice to tank-mix ethephon with defoliants. Even with this tank-mix approach, you can achieve improved boll opening and obtain generally good results.

Mepiquat chloride

Mepiquat chloride, marketed as Pix and other related products, controls plant height and cell elongation by suppressing the production of gibberellic acid. By controlling excessive growth, mepiquat chloride allows the plants to reallocate carbohydrates for use in formation and improved retention of cotton fruiting structures (squares and bolls). You can reduce the amount of mepiquat chloride applied during years with adequate heat units and good boll loads. A heavy fruit load will effectively keep plant heights at desirable levels, and can be combined with moderation in irrigation and nitrogen fertilization practices to prevent excessive vegetative growth.

Growing seasons with the fewest heat units (degree-days) resulted in the greatest benefits from the use of mepiquat chloride. Lint yield increases of more than 100 pounds per acre have been measured in years with cooler-than-normal summers. High rates of mepiquat chloride applied during warm years with adequate heat have sometimes resulted in yield responses that were slightly negative. Research has determined that cotton plants that are 30 inches or shorter at first bloom should not receive mepiquat chloride applications and cotton plants taller than 30 inches at first bloom would benefit most.

Research has shown that under some conditions, low-rate multiple applications of mepiquat chloride can result in even greater increases in lint yield. The first application should be made at first square, with subsequent applications at first bloom and again at mid-bloom. Best results with multiple low-rate applications occurred with narrow-row (30-inch row spacing) cotton during years with fewer heat units than normal, and better results were observed in the northern San Joaquin Valley than in other areas.

There is a limited amount of mepiquat chloride use late in the season, at or near vegetative cutout or around the first week in August. It is too late in the season then for the crop to reap the early season benefits of controlling plant height, setting squares, or retaining fruit. Benefits from late-season applications are more difficult to measure and include even crop maturity, easier defoliation, and fewer "buggy whips," or tall plants rising above the crop canopy.

Differences in mepiquat chloride use in 'Pima' cotton reflect that 'Pima' is a different species than the 'Acala' varieties of upland cotton that predominated in California until the 1990s. 'Pima' varieties have a longer growing season requirement, more fruiting sites, and more difficulty with defoliation. University of California research has generally recommended higher application rates for 'Pima' cotton and multiple applications made at later growth stages than for 'Acala' cottons. Results have shown less consistent yield responses for 'Pima,' but good utility in vegetative growth control.

More extensive recommendations and research results for cotton are found in the *Cotton Production Manual,* UC ANR Publication 3352.

References

Constable, G. 1996. Management of cotton with nitrogen and Pix. NSW Agriculture and Fisheries, Agricultural Research Station, Narrabri.

Kerby, T. A., et al. 1982. Effect of Pix on earliness, yield, and cotton plant growth when used at various nitrogen levels. Beltwide Cotton Conf. Proc., Memphis, Tennessee.

Kerby, T. A., B. L. Weir, and M. P. Keeley. 1996. The uses of Pix. Chapter 21 in Cotton production manual, S. J. Hake, T. A. Kerby, and K. D. Hake. Oakland: University of California, Division of Agriculture and Natural Resources, Publication 3352.

Weir, B. L. 1993. Second year trials show foliar nitrogen plus Pix can pay off on cotton. Reno: Gibbs and Soell.

Weir, B. L., et al. 1990. Sequential low-dose application of Pix: A four-year summary. Beltwide Cotton Conf. Proc., Memphis, Tennessee.

Weir, B. L., T. A. Kerby, R. N. Vargas, B. A. Roberts, L. L. Ede, K. Hake, and S. Johnson. 1991. Sequential low-dose applications of Pix: A four-year summary. Proc. Beltwide Cotton Conf. Vol. 2, pp. 1017–1018.

Weir, B., and D. Wiley. 1994. Evaluations of various formulations of Pix. Beltwide Cotton Conf. Proc., Memphis, Tennessee.

Rice

Gibberellic acid (GA)

Gibberellic acid (GA) is currently registered for use on rice in California, but very little is being used. In the mid-South states of the United States, GA seed-treated rice is used primarily when drill-seeding to improve seedling emergence. Rice emergence through the soil is not generally a problem in California, however, because most of the rice is water-seeded on the surface of flooded soils. If California production practices change in the future, this PGR may be utilized more in California rice production.

References

Carlson, R. D., N. Chang, C. B. Schaefer, and J. A. Fugiel. 1992. Efficacy of release seed treatment in California rice production systems. Proc. West. Plant Growth Regulator Soc.

Dunand, R. T. 1997. Post-germination seed treatment with gibberellic acid for water-seeded rice. Proc. West. Plant Growth Regulator Soc.

Dunand, R. T. 1998. Influence of gibberellic acid seed treatment timing and micronutrient interaction in water-seeded rice. Proc. West. Plant Growth Regulator Soc.

Wick, C. M., J. C. Neville, K. S. McKenzie, and J. R. Webster. 1992. How gibberellic acid is used in California rice. Proc. West. Plant Growth Regulator Soc.

6

Plant Growth Regulators for Citrus

CHARLES W. COGGINS, JR., Professor of Plant Physiology and Plant Physiologist Emeritus,
Department of Botany and Plant Sciences, University of California, Riverside

Note to reader: **This chapter contains adequate information for the citrus portion of the DPR exam. To be successful in your practice, you should consult current product labels (available online at http://www.cdms.net) and the *UC IPM Pest Management Guidelines for Citrus* (http://www.ipm.ucdavis.edu). These contain up-to-date details. For a sense of the extent of plant regulator use worldwide, see the article cited below.**

El-Otmani, M., C. W. Coggins, Jr., M. Agusti, and C. J. Lovatt. 2000. Plant growth regulators in citriculture: World current uses. Crit. Rev. Plant Sci. 19(5):395–447.

GENERAL INFORMATION

The plant growth regulators 2,4-dichlorophenoxyacetic acid (2,4-D), gibberellic acid (GA_3), and naphthaleneacetic acid (NAA) are registered for preharvest use on California citrus crops. 2,4-D is used mainly to delay and reduce unwanted abscission (fruit drop), GA_3 is used mainly to delay senescence (overripening), and NAA is used to promote abscission of excess fruit (thinning).

In order to be effective, plant growth regulators (PGRs) must be absorbed by plant tissue. Good spray coverage is essential and climatic conditions that favor absorption (warm and humid conditions) are therefore desirable. Consider such factors as tree size, canopy density, location of fruit, and type of spray equipment when deciding how much spray material will be required to achieve good coverage. Apply all spray materials uniformly to the fruiting canopy.

When you use 2,4-D to reduce drop of mature fruit, apply the compound before (preferably *shortly* before) fruit drop becomes a problem, but far enough ahead of flowering to minimize undesirable effects that 2,4-D would otherwise have on the spring cycle of growth. For navel oranges, October through December sprays are common and are generally effective. October may be too early to provide good control, however, when conditions favor fruit drop (e.g., warm winter, protracted harvest). Still, January sprays may be somewhat risky, especially when environmental factors favor an earlier-than-usual spring flush of growth.

For mature grapefruit and 'Valencia' orange trees, 2,4-D can be applied to control drop of mature fruit or as a dual-purpose spray (to control mature fruit drop and to improve fruit size for the next year's crop). For further explanation see the 'Valencia' oranges and grapefruit section, below. Fruit sizing sprays require excellent coverage. In general, 'Valencia' orange is more responsive than grapefruit to fruit-sizing sprays.

The purpose of applying GA_3 to citrus trees in California is to delay fruit senescence. Applications should be made while the fruit are still physiologically young, but are approaching maturity. For further discussion, see the section on Preharvest Uses, below. GA_3 can have a negative effect on flowering and thus on production for the following year, especially if it is applied much later than specified on the current label or in the *UC IPM Pest Management Guidelines for Citrus.* It delays changes in rind color, an effect that can be considered either desirable or undesirable. For example, you apply GA_3 to navel orange trees while the fruit still have green rinds, delayed coloring will have a negative effect on your ability to harvest and market the fruit early in the season. Such an effect is desirable for late-harvested fruit, however, since it also delays rind senescence, yielding fruit that are paler in color than the deeper-colored fruit from untreated trees. GA_3 applications amplify the re-greening of 'Valencia' oranges. This is considered undesirable and can be minimized if you apply the compound no later than the date specified on the label or in the *UC IPM Pest Management Guidelines for Citrus.* GA_3 application usually causes leaf drop, which can be severe, especially when it is applied to navel orange trees that are under heat or water stress. When this happens, the tree may also suffer twig die-back. By including 2,4-D in the GA_3 spray, you may be able to reduce this kind of damage. These negative effects argue against the practice of applying GA_3 to young citrus trees.

NAA can be a very effective fruit-thinning agent. See the Fruit Thinning section (below) for information on its potential usefulness. Also, see the item on NAA in the Precautions list that follows.

2,4-D and GA_3 seem to be compatible with urea, potassium foliar sprays, zinc and manganese micronutrient sprays, and neutral copper sprays, but the timing of growth regulator applications may not coïncide with the best time for nutrient sprays.

Precautions When Using Growth Regulators:

- Plant growth regulators are potent compounds; care is warranted in their use.
- Avoid 2,4-D spray drift to susceptible plants, which include cotton, grapes, roses, beans, peas, alfalfa, lettuce, ornamentals, and all broadleaf species.
- If 2,4-D is applied shortly before or during a flush of growth, vegetative and reproductive growth may be damaged. This may result in lower fruit production, especially if the spring flush in affected.
- The effectiveness of 2,4-D for controlling fruit drop is enhanced by oil and decreased by calcium hydroxide (calcium hydrate, hydrated lime). The magnitude of the oil enhancement and the magnitude of the reduction caused by calcium hydroxide are not sufficiently understood to permit any extrapolation of University of California recommendations or product label instructions. Do not vary from label rates.
- GA_3 is slowly hydrolyzed by water and rapidly converted into an inactive isomer in highly alkaline solutions. Protect liquid and powder formulations from moisture and do not add GA_3 to highly alkaline spray mixtures. Although GA_3 seems to be stable in solutions of up to pH 11 for short periods of time (2 hours), its activity diminishes rapidly at the high pH values found in Bordeaux and whitewash mixtures. As a general rule, do not expose GA_3 to solutions higher than pH 8. Values below pH 8 may provide greater stability for GA_3 and better absorption by plant tissue.
- NAA is registered as a fruit-thinning agent for certain types of citrus. Label rates are 100 to 500 ppm. Within this concentration range, an application may still cause anywhere from inadequate to excessive thinning. In general, inadequate thinning occurs from the lowest label rate when maximum daytime temperatures on the day of application and several days thereafter are relatively low (say 85°F [29°C]). Excessive thinning generally occurs from the highest label rate when maximum daytime temperatures on the day of application and several days thereafter are relatively high (say 100°F [38°C]). In addition, excessive thinning can occur when NAA is applied to unhealthy or water-stressed trees.
- Older recommendations and product labels specified 2,4-D and GA_3 dosages in terms of concentrations (ppm or mg/liter). Because current spray volumes vary widely, University of California recommendations and current labels specify the amount of product per acre rather than ppm or mg/liter. If applied properly (i.e., if coverage is adequate and if the spray deposit does not dry rapidly), an application of a particular per-acre dosage of 2,4-D or GA_3 will be effective for the control of mature fruit drop and for delayed fruit senescence when applied at spray volumes of 100 to 750 gallons per acre. Lower-volume applications are less forgiving of imprecise spraying than are higher-volume applications. Success with 2,4-D applications to improve fruit size and with NAA as a fruit-thinning agent requires excellent coverage and wetting. As far as this author is aware, success with neither has been demonstrated when low-volume equipment has been used.
- In general, surfactants (wetting agents) help achieve good spray coverage. Many surfactant formulations are available in the marketplace. Some can cause rind blemishes on citrus fruit, so you need to find a suitable surfactant for citrus, whether through direct experimentation or by contacting an experienced citrus pest control operator. Research has shown that a suitable organosilicone adjuvant such as Silwet L-77 can increase the efficacy of GA_3 applied to navel orange trees. Two cautions are in order: increased efficacy also means increased risks of negative effects, such as excessive leaf drop and twig die-back from GA_3; and rind blemishes have been reported from relatively high adjuvant concentrations (a concentration of 0.025%, v/v [active ingredient basis] has a good rind blemish safety record).

PREHARVEST USES

Fruit Set

For many years, GA_3 has been used in Florida and in many countries as a fruit setting agent for self-incompatible mandarins and mandarin hybrids. Through a Special Local Need Registration, it became available in 2000 as a PGR to increase fruit set in Clementine mandarins in California.

Fruit Thinning

Some mandarin cultivars, mandarin hybrids, and (in some areas) 'Valencia' oranges have strong alternate bearing tendencies. You can help minimize this by thinning with NAA at an early stage of fruit development at the start of a heavy crop year. If you can obtain the desired degree of thinning, the remaining fruit may be larger at maturity, and if total yield is sufficiently reduced, the tree will flower and produce a good crop in the following "off-crop" year. The amount of thinning you obtain with NAA can be highly variable. See the item on NAA in the Precautions list, above.

Reference

Hield, H. Z., and R. H. Hilgeman. 1969. Alternate bearing and chemical fruit thinning of certain citrus varieties. Proc. of First Int'l. Citrus Symp. 3:1145–53.

Reducing Rind Senescence and Drop of Mature Fruit

Navel oranges

The softening of the rind on navel oranges, particularly late in the harvest season, can cause marketing problems. Sprays of GA_3 will delay the aging and accompanying softening of the rind (rind staining, water spot, and sticky rind). Internal fruit quality is not greatly influenced by GA_3 treatment.

GA_3 is commonly applied to orchards that are to be harvested in the second half of the season. Water sprays containing GA_3 are applied in September or October. Where late coloring is expected or where only a minimum delay in coloring can be tolerated, a later spray time based on rind color has been used. January applications are avoided, since they may decrease the subsequent year's production. A combination GA_3 and 2,4-D spray application is common practice, although the optimal times of application for the two PGRs do not coincide. The effectiveness of 2,4-D is decreased and the effectiveness of GA_3 is eliminated when the spray mixture contains whitewash (lime).

2,4-D can be used as a PGR to control preharvest mature fruit drop. Navel oranges are stored on the tree during a harvest season that may be 6 months long. Late in the season, fruit drop is common. Water spray applications of 2,4-D will reduce fruit drop by about half.

You can increase the fruit size of navel oranges at maturity by applying 2,4-D water sprays over the young fruit a few weeks after petal fall. This treatment will increase the fruit size by approximately one commercial size. There is a slightly disproportionate increase in rind thickness and an increase in rind roughness. This application should not be made in orchards that usually produce coarse-textured fruit, since the grade could be reduced due to excessive rind roughness. The dosage of 2,4-D depends on the size of the young fruit at the time of treatment. Careful timing of the sprays is required to match average fruit size with appropriate 2,4-D dosage and also to avoid application during a period of vegetative growth. You should not apply 2,4-D to trees less than 6 years old. Specific details on the fruit sizing application of 2,4-D can be found in the plant growth regulators section of the *UC IPM Pest Management Guidelines for Citrus* (see References).

References

Coggins, C. W., Jr. 2001. Plant growth regulators, in: UC IPM pest management guidelines: Citrus. Oakland: University of California, Division of Agriculture and Natural Resources, Publication 3339.

Coggins, C. W., Jr., and H. Z. Hield. 1968. In: W. Reuther, L. D. Batchelor, and H. J. Webber (eds.), The citrus industry, Vol. II, pp. 371–389. Berkeley: University of California, Division of Agricultural Sciences.

Hield, H. Z., R. M. Burns, and C. W. Coggins, Jr. 1964. Preharvest use of 2,4-D on citrus. Berkeley: University of California, Division of Agricultural Sciences, Leaflet 2447.

'Minneola' tangelo

Rind senescence and fruit abscission are problems when you harvest this cultivar late in the season. As with navel orange cultivars, you can reduce these problems through timely applications of GA_3 and 2,4-D. The production of mandarins and mandarin hybrids is likely to increase in California, and GA_3 and 2,4-D probably will be registered for similar uses on this general category of citrus.

Reference

Coggins, C. W., Jr. 2001. Plant growth regulators, in: UC IPM pest management guidelines: Citrus. Oakland: University of California, Division of Agriculture and Natural Resources, Publication 3339.

'Valencia' oranges and grapefruit

'Valencia' oranges and grapefruit are similar to one another in that they often have two crops of fruit on the tree in the early summer. An adaptation of 2,4-D spray applications can permit a single application to both increase the mature size of the young fruit and reduce the fruit drop of the older crop that is still on the tree. The dosage of such a spray application is based on the average size of the young fruit. The 2,4-D dosage increases as the fruit diameter increases. The greatest diameter at which the young fruit are responsive is 1 inch for grapefruit and ¾ inch for 'Valencia' oranges. The single spray application should be timed to avoid an active period of leaf development.

You can also apply 2,4-D only to reduce mature fruit drop without influencing the later size of the young fruit. In this instance, apply a spray of 2,4-D after young 'Valencia' orange fruit are at least ¾ inch in diameter or after young grapefruit are at least 1 inch in diameter. Harvest of mature fruit already on the tree must be delayed 7 days after treatment, and spray should not be applied during periods of soft vegetative growth or on trees younger than 6 years old.

Timely sprays of GA_3 will delay aging of the rind on grapefruit and 'Valencia' oranges. At this writing, GA_3 is registered for use in 'Valencia' orange groves in California, but its use on grapefruit is restricted to Florida and Texas. The California restriction regarding grapefruit may be removed in the near future.

GA_3 is used to delay 'Valencia' creasing (albedo breakdown) and to delay aging and softening of the rind. To obtain this response and to reduce undesirable effects on rind color, apply GA_3 at an early stage of fruit development (typically in August or September when target crop fruit are small). Application at a later stage may lead to more delay in coloring or greater re-greening than is acceptable. Also, some re-greening may occur on mature fruit that are present on the tree at application time. Relatively few 'Valencia' orange groves receive GA_3 treatment, probably because of these negative color responses and the fact that GA_3 reduces rather than eliminates albedo breakdown.

Apply GA_3 to grapefruit groves when fruit are approaching color break or as soon as they achieve marketable color, subsequent softening of the rind is delayed to a sufficient degree to prolong the harvest season. Under California conditions, such sprays can cause a superficial rind blemish, especially if a surfactant is used and especially if the application is made to fruit that have not yet reached marketable color. This accounts for the current California restriction.

Because some California grapefruit growers and packers want access to GA_3 in spite of these negative factors, it is likely that the California restriction will be lifted and that suitable caution statements will be added to product labels. Because GA_3 may reduce the accumulation of lycopene (a pink pigment), it is likely that GA_3 will be used more on white-fleshed rather than on pigmented grapefruit.

References

Coggins, C. W., Jr. 2001. Plant growth regulators, in: UC IPM pest management guidelines: Citrus. Oakland: University of California, Division of Agriculture and Natural Resources, Publication 3339.

Hield, H. Z., R. M. Burns, and C. W. Coggins, Jr. 1964. Preharvest use of 2,4-D on citrus. Berkeley: University of California, Division of Agricultural Sciences, Leaflet 2447.

Lemons

Growers determine whether lemons are ready for harvest based on fruit size or well-developed rind color. Large, green fruit are picked and stored until they develop enough color for marketing. Yellow lemons are picked and either marketed as fresh fruit or processed. The greatest demand for fresh lemons is during the summer, and the peak of production is in the spring.

Applications of GA_3 when the target crop's fruit are still green and one-half to three-quarters of their full size will delay lemon fruit maturity. A GA_3 application may reduce bloom and fruit set the spring after treatment. It may also increase fruit set during the following summer. A GA_3 application can delay the harvest date for fruit in the current year by delaying fruit maturity, and can shift the production peak for the year following treatment.

You can reduce drop of mature lemon fruit by applying 2,4-D water sprays during October through December. Later applications could cause misshapen leaves for the spring growth flush. 2,4-D should not be applied during an active period of vegetative growth or to trees less than 6 years old.

References

Coggins, C. W., Jr. 2001. Plant growth regulators, in: UC IPM pest management guidelines: Citrus. Oakland: University of California, Division of Agriculture and Natural Resources, Publication 3339.

Coggins, C. W., Jr., and H. Z. Hield. 1968. In: W. Reuther, L. D. Batchelor, and H. J. Webber (eds.). The citrus industry Vol. II, pp. 371–389. Berkeley: University of California, Division of Agricultural Sciences.

Hield, H. Z., R. M. Burns, and C. W. Coggins, Jr. 1964. Preharvest use of 2 4-D on citrus. Berkeley: University of California, Division of Agricultural Sciences, Leaflet 2447.

Limes

Green colored lime fruit are desired in the marketplace, and you can use GA_3 to delay rind coloration and keep them green. To obtain this response, apply GA_3 when the target crop is one-half to three-quarters of its full size and still green.

Reference

Coggins, C. W., Jr. 2001. Plant growth regulators, in: UC IPM pest management guidelines: Citrus. Oakland: University of California, Division of Agriculture and Natural Resources, Publication 3339.

Reduction of Leaf and Fruit Drop Following Pesticide Oil Sprays

Navel oranges, lemons, grapefruits, and 'Valencia' oranges

Pesticide oil sprays are applied to citrus at times of the year when the applications may be followed by hot, dry winds. Such conditions may lead to leaf and fruit drop.

The addition of the isopropyl ester of 2,4-D reduces this oil-induced leaf and fruit drop. The ester form of 2,4-D is oil-soluble and oil enhances the entry of 2,4-D into the plant. For this reason, you should not apply 2,4-D in oil or in water solutions containing high surfactant levels (which can have the same effect) at the dosages recommended for water sprays.

Older recommendations and product labels called for the addition of the isopropyl ester of 2,4-D to the pesticide oil spray mixture at a concentration of 4 ppm (acid equivalent), based on total spray volume (high-volume spraying was standard). Because current spray volumes vary widely and because the isopropyl ester of 2,4-D partitions into the oil phase of the oil-in-water emulsion, current recommendations and product labels call for the addition of the isopropyl ester of 2,4-D at the rate of 2.2 ml of formulation per gallon of oil (0.88 grams acid equivalent). 2,4-D should not be applied to trees younger than 6 years old or during a growth flush to trees of any age.

References

Coggins, C. W., Jr. 2001. Plant growth regulators, in: UC IPM pest management guidelines: Citrus. Oakland: University of California, Division of Agriculture and Natural Resources, Publication 3339.

Coggins, C. W., Jr., and H. Z. Hield. 1968. In: W. Reuther, L. D. Batchelor, and H. J. Webber (eds.). The citrus industry Vol. II, pp. 371–389. Berkeley: University of California, Division of Agricultural Sciences.

Hield, H. Z., R. M. Burns, and C. W. Coggins, Jr. 1964. Preharvest use of 2,4-D on citrus. Berkeley: University of California, Division of Agricultural Sciences, Leaflet 2447.

POSTHARVEST USES

Lemons

2,4-D

Lemons differ from other citrus in that they may be subjected to lengthy packinghouse storage. The postharvest application of 2,4-D can increase the vitality of lemons and, in so doing, indirectly reduce their susceptibility to decay. The isopropyl ester form of 2,4-D is applied as a final step in the washing or waxing procedure just prior to storage. This treatment will slightly delay the loss of chlorophyll. The major benefit is the resulting increased vigor and persistence of the button (clipped stem end-calyx attachment). Retention of the button retards entry of *Alternaria* fungi through the tissue exposed when the button detaches.

References

Coggins, C. W., Jr., and H. Z. Hield. 1968. In: W. Reuther, L. D. Batchelor, and H. J. Webber (eds.). The citrus industry Vol. II, pp. 371–389. Berkeley: University of California, Division of Agricultural Sciences.

DeWolfe, T. A., L. C. Erickson, and B. L. Brannaman. 1959. Retardation of Alternaria rot in stored lemons with 2,4-D. Proc. Amer. Soc. Hort. Sci. 74:367–371.

Grapefruit, Lemons, Oranges, and Mandarins

Ethylene

Treatment with ethylene gas for de-greening (to remove green color from the rind) enhances yellow or orange rind color for sales appeal. The fruit must be far enough advanced in maturity for the rind to contain yellow or orange pigments. During de-greening, ethylene treatment removes the green chlorophyll from the rind and the yellow or orange pigments become more prominent. De-greening treatments usually are made at the start of a harvest season when the yellow or orange rind color has not fully developed or later in the harvest season for re-greened 'Valencia' orange or grapefruit.

For de-greening, fruit are placed in enclosed rooms with controlled temperature (65° to 75°F [18° to 24°C]) and relative humidity (85 to 98 percent). The ethylene is metered into storage spaces to maintain atmospheric concentrations of 1 to 10 ppm. The concentration used depends on the type of citrus being de-greened, the temperature, and the amount of de-greening needed. Temperature depends on the type and condition of the fruit. Relative humidity (RH) depends on the type of citrus being de-greened and on the stage of fruit development. Here are a few examples for California citrus (note that these are only examples, and not specific recommendations for use):

Turgid navel orange fruit early in the harvest season: 68° to 72°F (20° to 22°C), 85 to 90 percent RH, 2.5 to 5 ppm ethylene, and a de-greening time of 5 to 7 days. As the navel orange harvest season progresses, but before de-greening becomes unnecessary, the time for adequate de-greening will decline to as few as 2 days.

Late-season 'Valencia' oranges: 72°F (22°C), 95 to 98 percent RH, 10 ppm ethylene, and a de-greening time of 2 to 4 days.

The most common negative effect of de-greening conducted under desirable conditions is an increase in decay due to an increase in the postharvest time during which fruit are not protected by postharvest fungicides. For this reason, export markets, which require more transport time than domestic markets, are best served by fruit that require only short de-greening times. Because improper de-greening conditions can cause additional problems such as rind disorders, black buttons, and accelerated rind aging, the specifications for any given de-greening treatment should be set by a knowledgeable person. Ethylene gas is explosive in concentrations from 3 to 34 percent in air. If you take due care, there is little hazard of explosion during de-greening, since the concentration required for de-greening is so low and the low concentration is kept uniform throughout the treatment chamber by proper air circulation and ventilation.

Reference

Grierson, W., E. Cohen, and H. Kitagawa. 1986. Degreening. In: W. F. S. Wardowski, S. Nagy, and W. Grierson (eds.). Fresh citrus fruits, pp. 253–274. Westport, CT: AVI Publishing Co., Inc.

Citrus

GA_3

When GA_3 is applied in storage wax to lemons, it delays senescence, which maintains the fruit's natural resistance to sour rot (*Geotrichum candidum* Link ex Pers.) and otherwise allows for a longer storage life. When GA_3 is added to the shipping wax applied to yellow lemons and other mature citrus fruits, it can often delay rind senescence, including rind color changes. The resulting delay in postharvest coloring (de-greening) of re-greened 'Valencia' oranges is undesirable, but delays in the coloring of other citrus fruits are considered to be beneficial.

Reference

Coggins, C. W., Jr., M. F. Anthony, and R. Fritts, Jr. 1992. The postharvest use of gibberellic acid on lemons. Proc. Intl. Soc. Citriculture. pp. 478–481.

PROMOTION OF ROOTING

Grapefruit, Lemons, Oranges, and Mandarins

Commercial citrus plantings generally are made by budding selected scions (tops) onto seedling rootstocks that are selected for resistance to *Phytophthora* spp., nematodes, and soil salt accumulations. You can propagate citrus by rooting cuttings. Lemons, limes, and citrons root easily. Mandarins are difficult to root. Indolebutyric acid (IBA) is a commonly used rooting aid. Other rooting aids that have shown success on certain cultivars are naphthaleneacetic acid (NAA) and indoleacetic acid (IAA). Techniques for rooting citrus and using rooted cuttings for commercial progagation can be found in the references listed for this section.

References

Erickson, L. C., and W. P. Bitters. 1954. Effects of various plant growth regulators on rooting of cuttings of citrus and related species. Proc. Amer. Soc. Hort. Sci. 61:84–88.

Halma, F. F. 1947. Own-rooted and budded lemon trees. Calif. Citrog. 33:2–3, 14–15.

Optiz, K. W., R. G. Platt, and E. F. Frolich. 1968. Propagation of citrus. Berkeley: University of California Division of Agricultural Sciences, Circular 546.

Platt, R. G., and K. W. Optiz. 1973. In: W. Reuther (ed.). The citrus industry Vol. III, pp. 1–47. Berkeley: University of California Division of Agricultural Sciences.

7

Plant Growth Regulators for Ornamental Plants and Turfgrass

GARY W. HICKMAN, Horticulture Advisor, Mariposa County,
ED PERRY, Farm Advisor, Stanislaus County, and PAM GEISEL, Farm Advisor, Fresno County

Plant growth regulators (PGRs) are used on a variety of ornamental plants and turfgrasses to control height, prevent fruit set, reduce basal trunk sprouting, and increase the rooting of cuttings. They are also used to induce flowering and increase lateral branching on some plants.

Unlike some other chemicals such as insecticides or fungicides, many PGRs produce different responses depending on what species or even cultivar of ornamental plant is treated. The timing of application, dosage, and need for repeat applications must be determined separately for each species treated. Rates that are too low or too high or applications that are improperly timed for the species or cultivar may fail to yield an effective treatment or may even cause a phytotoxic reaction. Adverse environmental or cultural conditions that cause plant stress can also affect the efficacy of PGRs.

Gibberellin Biosynthesis Inhibitors

Ancymidol, chlormequat, flurprimidol, paclobutrozol, uniconazole

By partially inhibiting the synthesis of gibberellin in plants, some gibberellin biosynthesis inhibitors cause internode compression in a wide variety of ornamental plants. The primary method of plant uptake is through the xylem, so soil or stem applications are usually the most effective. The target plants' developmental stage at time of treatment has a substantial influence on efficacy of treatments. Effective rates vary depending on species, cultivar, climate, and growth stage.

Ancymidol – Greenhouse and nursery plants. Low pH in the media reduces PGR activity. May delay flowering.

Chlormequat – Greenhouse and nursery plants. Inhibits the elongation of cells. Applied as either a spray or drench. Spray treatments can cause leaves to yellow.

Flurprimidol – Ornamental trees and turfgrass. In turfgrass, more actively absorbed by the roots, so it should be watered in after application.

Paclobutrozol – Greenhouse and nursery plants, ornamental trees, and turfgrass. High application rates may delay the flowering of herbaceous plants.

Uniconizole – Greenhouse and nursery plants and ornamental trees. PGR activity is reduced when applied to media containing pine bark.

Terminal Bud Inhibitors

NAA, dikegulac-sodium, maleic hydrazide, mefluidide

By inhibiting or killing apical meristematic tissue, terminal bud inhibitors reduce terminal growth and may stimulate growth of lateral buds.

Naphthaleneacetic acid (NAA) – Applied after pruning, NAA inhibits the basal sprouting of woody ornamentals. Promotes rooting of some plants. Prevents the formation of unwanted fruit on some ornamental trees.

Dikegulac-sodium – Reduces or eliminates bloom and prevents fruit set on landscape trees; also used as a chemical pinching agent.

Maleic hydrazide – Inhibits cell division, but not cell elongation, in the meristem. Retards terminal growth and increases lateral growth of ornamental trees and shrubs. Also used to reduce the rate of turfgrass growth. May cause temporary yellowing in some turfgrass species.

Mefluidide – Used on turfgrass and woody ornamentals. Absorbed by leaf, not translocated. This PGR is used on turf to reduce its growth and suppress the formation of seed and on woody ornamentals to reduce their growth. Allow applied material to dry on the plant before irrigation or rain.

Others

Daminozide, ethephon

Daminozide – Greenhouse and nursery plants. Used for height control and for the promotion of flowering. Activity is decreased by high temperatures.

Ethephon – Landscape trees and nursery plants. Ethephon is registered for several uses on ornamentals. With proper application timing, it eliminates the formation of unwanted fruit on some ornamental trees. Application should be made from the flower bud stage to full bloom stage, before fruit set, for effective control. May cause partial, temporary defoliation in drought-stressed trees. Ethephon can also be used to stimulate the lateral branching of azaleas and geraniums and to control plant height in daffodils and hyacinths. It can promote earlier defoliation of roses and induce flowering in some bromeliad species. Ethephon is also used to control broadleaf mistletoes in ornamental trees.

Chemical	Trade name(s)
Ancymidol	A-Rest
Chlormequat	Cycocel
Flurprimidol	Cutless
Paclobutrazol	Bonzi
Uniconazole	Sumagic
NAA	Hormodin
Dikegulac-sodium	Atrimmec
Maleic hydrazide	Maintain
Mefluidide	Embark
Daminozide	B-Nine, Alar
Ethephon	Ethrel, Florel

Reference

Basra, A. S., ed. 2000. Plant growth regulators in agriculture and horticulture. New York: Food Products Press.

Glossary

Abscisic acid: A naturally occurring plant growth substance that tends to promote abscission and inhibit growth.

Abscission: The process by which organs (such as leaves or fruit) separate from a plant by way of breakage through a specialized region called the abscission layer.

Adjuvant: Any substance in a formulated product that improves or alters the characteristics of the formulation.

Adventitious: Referring to a structure that is growing in an uncharacteristic place on a plant, such as buds at other places than leaf axils, or roots growing from leaves or stems.

Aleurone layer: The top cell layer of the endosperm; composed of protein in grains such as wheat.

Apical dominance: The suppression by the terminal (apical) bud of lateral bud development.

Apical meristem: see *Meristem.*

Auxin: A class of naturally occurring plant growth substances essential to cell differentiation, division, and enlargement.

Avena coleoptile curvature test: A bioassay designed to measure the effects of plant growth regulators on the *coleoptile* (the protective sheath surrounding an active grass shoot tip). The amount of curving of the coleoptile is proportionate to the amount of plant growth regulator to which it has been exposed.

Axil: The upper angle formed by a leaf or branch as it grows from the main stem.

Axillary: A term applied to buds or branches located or growing from an axil.

Bioassay: A "living test" that usually is specific for a particular family of chemical compounds. The bioassay may consist of a test plant, a plant part, or even a microorganism.

Boll: The bud of a cotton plant, from which the cotton fiber is harvested.

Bolting: In some annual and biennial plant species (e.g., cole crops, lettuce), the change from vegetative growth to reproductive growth (flowering and seed set).

Calibration: The standardization or correction of measuring devices on instruments to match a common standard; the adjustment of nozzles on a spray rig to accurately dispense a given quantity or mode of chemical treatment.

Cambium: The layer of tissue between the bark and wood in woody plants, from which new wood and bark develops.

Cold hardiness: The ability of a plant to survive the extreme cold and drying effects of winter weather.

Cotyledon: A "seed leaf" of a broadleaf plant, which either stores or absorbs food. *Plural:* The first two leaves to emerge from a seed after germination.

Cultivar: A variety or race of plant that has originated and persisted under cultivation; not necessarily referable to a botanical species.

Cytokinesis: Cell division; the division of the cell's cytoplasm to produce two daughter cells, occurring soon after *mitosis.*

Cytokinins: A group of naturally occurring plant growth substances and synthetic analogs that are active in stimulating cell division and in delaying senescence of plant organs.

Dioecious: Of a plant species or variety; having the male and female reproductive structures on separate plants. Compare *Monoecious*.

Dormancy: A condition of arrested growth that occurs only after less-favorable environmental conditions; for example, low light intensity, short day lengths, or colder temperatures.

Drift: The dispersal of a substance, such as a pesticide or plant growth regulator, beyond the intended application area.

Emulsifying agent: A substance that can be used to produce an emulsion of two liquids that normally cannot be mixed together (such as oil and water).

Endogenous: Of internal origin, growing from within.

Endosperm: Nutritive tissue in a seed.

Epinasty: A plant's response to a condition (such as application of a plant growth regulator) that results in disturbance of the normal growth pattern and results in the curving, twisting, or cupping of affected plant parts.

Ethephon: The generic term for a synthetically produced ethylene-releasing compound, (2-chloroethyl) phosphonic acid.

Ethylene: A gaseous substance produced by all plant tissues, which seems to be active in the regulation of many plant processes.

Gibberellins: A family of compounds produced by plants and by the fungus *Gibberella fujikuroi*. They are important in a number of plant processes.

Gynoecious: Of an organism, variety, or species; something that lacks male reproductive structures, such as a plant with all female flowers.

Hypocotyl: The part of the stem of an embryo or young seedling that grows below the cotyledons.

Inflorescence: The group or arrangement in which flowers are borne on a plant.

Inhibitor: A class of plant growth substances that, at nontoxic concentrations, retard the normal pattern of growth. For example, elongation inhibitor, an inhibitor of growth from seed germination to flowering, or an inhibitor of a plant's responses to light.

Lateral: Arising from the side of or directed to the side. A lateral root or stem is one that arises from another, older root or stem. Also called secondary roots or branches.

Lateral meristem: see *Meristem* and *Lateral*.

Locules: Cavities of a plant's ovary in which ovules occur.

Meristem: Zone or localized region of active cell division in plants from which permanent tissue is derived. Meristems are the areas in plants where *mitosis* (cell division) occurs, and thus where growth occurs. *Lateral meristems* can be found where secondary tissue is growing (e.g., thickening of stems, roots, or cork), while the *apical meristems* responsible for vertical growth can be found at the root and shoot (apex) tips.

Mitosis: The process of nuclear division that produces two daughter cells from one mother cell, all of which are genetically identical.

Monoecious: Of a plant species or variety; having the male and female reproductive structures in separate flowers but on the same plant. Compare *Dioecious*.

Parthenocarpy: Development of a fruit without sexual fertilization.

Parthenogenesis: Development of an organism without sexual fertilization.

Parts per million (ppm): The number of molecules of a specified substance that are present in a volume of 1 million molecules of containing medium; a unit of concentration often used when measuring low levels of a substance.

Phenotype: The sum of the characteristics manifested by a given organism, as contrasted with the set of genes it possesses (its *genotype*).

Phloem: The tissue in the conducting system of a plant through which metabolites (products of chemical reactions in the plant, such as sugars) are transported.

Photoperiod: The number of hours of a day that the environment is light.

Phytohormones: see *Plant growth substances.*

Phytotoxicity: The quality of being poisonous to plants.

Plant growth inhibitor: see *Inhibitor.*

Plant growth regulators: Naturally occurring or synthetic organic compounds that in small amounts promote, inhibit, or otherwise modify a plant's physiological processes.

Plant growth retardant: see *Inhibitor.*

Plant growth substances: Naturally occurring organic compounds that are active at low concentrations and are translocated from a site of production to a site of action, and that in small amounts promote, inhibit, or otherwise modify a plant's physiological processes.

Plant hormones: see *Plant growth substances.*

Rachis: The axis of an *Inflorescence.*

Rest period: A condition in which a plant's growth is arrested, even when in the presence of favorable environmental conditions.

Senescence: The process of aging and loss of function of cells, organs, or intact plants. An example is the normal yellowing of a leaf.

Stem cutting: A piece of plant stem that is cut from the plant and used to vegetatively propagate a plant.

Surfactant: A surface-active agent; an adjuvant used to improve a pesticide's ability to stick to and to be absorbed by the target surface. Also known as a *wetting agent.*

Tank mix: A mixture of two or more pesticides or fertilizers and pesticides applied at the same time through the same application equipment.

Thinning: The partial or complete removal of undesirable or excess plants or plant parts.

Vascular cambium: The lateral meristem that forms the secondary tissue and is located between the *xylem* and the *phloem.*

Wetting agent: see *Surfactant.*

Xylem: The tissue in vascular plants that carries water and nutrients up from the roots.

CPSIA information can be obtained
at www.ICGtesting.com
Printed in the USA
LVHW070125070319
609772LV00004B/38/P

9 781601 074188